The Theory of Quantum Liquids

Superfluid Bose Liquids

VOLUME II

The Theory of Quantum Liquids Superfluid Bose Liquids

PHILIPPE NOZIÈRES
Institut Max Von Laue-Paul Langevin

DAVID PINES
University of Illinois

Advanced Book Program

CRC Press
Taylor & Francis Group
Boca Raton London New York

CRC Press is an imprint of the
Taylor & Francis Group, an **informa** business

The Theory of Quantum Liquids, Volume II: Superfluid Bose Liquids

First published 1990 by Westview Press

Published 2018 by CRC Press
Taylor & Francis Group
6000 Broken Sound Parkway NW, Suite 300
Boca Raton, FL 33487-2742

ISBN 13: 978-0-201-50063-9 (hbk)
ISBN 13: 978-0-201-40841-6 (pbk)

Visit the Taylor & Francis Web site at
http://www.taylorandfrancis.com

and the CRC Press Web site at
http://www.crcpress.com

Library of Congress Cataloging-in-Publication Data

Nozières, P. (Philippe)
Theory of quantum liquids. vol 2. superfluid bose liquids /
Philippe Nozières and David Pines
(Advanced book classics series)
Pines' name appears first on vol. 1.
Includes bibliographies and index.

1. Quantum liquids. I. Pines, David. II. Title. III. Series.
QC174.4.N69 1990 530.1'33–dc19 90-24104

This unique image was created with special-effects photography. Photographs of a broken road, an office building, and a rusted object were superimposed to achieve the effect of a faceted pyramid on a futuristic plain. It originally appeared in a slide show called "Fossils of the Cyborg: From the Ancient to the Future," produced by Synapse Productions, San Francisco. Because this image evokes a fusion of classicism and dynamism, the future and the past, it was chosen as the logo for the Advanced Book Classics series.

Publisher's Foreword

"Advanced Book Classics" is a reprint series which has come into being as a direct result of public demand for the individual volumes in this program. That was our initial criterion for launching the series. Additional criteria for selection of a book's inclusion in the series include:

- Its intrinsic value for the current scholarly buyer. It is not enough for the book to have some historic significance, but rather it must have a timeless quality attached to its content, as well. In a word, "uniqueness."
- The book's global appeal. A survey of our international markets revealed that readers of these volumes comprise a boundaryless, worldwide audience.
- The copyright date and imprint status of the book. Titles in the program are frequently fifteen to twenty years old. Many have gone out of print, some are about to go out of print. Our aim is to sustain the lifespan of these very special volumes.

We have devised an attractive design and trim-size for the "ABC" titles, giving the series a striking appearance, while lending the individual titles unifying identity as part of the "Advanced Book Classics" program. Since "classic" books demand a long-lasting binding, we have made them available in hardcover at an affordable price. We envision them being purchased by individuals for reference and research use, and for personal and public libraries. We also foresee their use as primary and recommended course materials for university level courses in the appropriate subject area.

The "Advanced Book Classics" program is not static. Titles will continue to be added to the series in ensuing years as works meet the criteria for inclusion which we've imposed. As the series grows, we naturally anticipate our book buying audience to grow with it. We welcome your support and your suggestions concerning future volumes in the program and invite you to communicate directly with us.

Advanced Book Classics

V.I. Arnold and A. Avez, *Ergodic Problems of Classical Mechanics*

E. Artin and J. Tate, *Class Field Theory*

Michael F. Atiyah, *K-Theory*

David Bohm, *The Special Theory of Relativity*

P.C. Clemmow and J. P. Dougherty, *Electrodynamics of Particles and Plasmas*

Ronald C. Davidson, *Theory of Nonneutral Plasmas*

P.G. deGennes, *Superconductivity of Metals and Alloys*

Bernard d'Espagnat, *Conceptual Foundations of Quantum Mechanics, 2nd Edition*

Richard Feynman, *Photon-Hadron Interactions*

Dieter Forster, *Hydrodynamic Fluctuations, Broken Symmetry, and Correlation Functions*

William Fulton, *Algebraic Curves: An Introduction to Algebraic Geometry*

Kurt Gottfried, *Quantum Mechanics*

Leo Kadanoff and Gordon Baym, *Quantum Statistical Mechanics*

I.M. Khalatnikov, *An Introduction to the Theory of Superfluidity*

George W. Mackey, *Unitary Group Representations in Physics, Probability and Number Theory*

A. B. Migdal, *Qualitative Methods in Quantum Theory*

Philippe Nozières and David Pines, *The Theory of Quantum Liquids, Volume II : Superfluid Bose Liquids*- new material, 1990 copyright

David Pines and Philippe Nozières, *The Theory of Quantum Liquids, Volume I: Normal Fermi Liquids*

F. Rohrlich, *Classical Charged Particles - Foundations of Their Theory*

David Ruelle, *Statistical Mechanics: Rigorous Results*

Julian Schwinger, *Particles, Source and Fields, Volume I*

Julian Schwinger, *Particles, Sources and Fields, Volume II*

Julian Schwinger, *Particles, Sources and Fields, Volume III* - new material, 1989 copyright

Jean-Pierre Serre, *Abelian ℓ-Adic Representations and Elliptic Curves*

R.F. Streater and A.S. Wightman, *PCT Spin and Statistics and All That*

Rene Thom, *Structural Stability and Morphogenesis*

Vita

Philippe Nozières
Professor of Physics at the Collège de France, Paris, he studied at the Ecole Normale Superiéure in Paris and conducted research at Princeton University. Dr. Nozières has served as a Professor at the University of Paris and at the University of Grenoble. His research is currently based at the Laue Langevin Institute in Grenoble. A member of the Académie des Sciences, he has been awarded the Wolf Prize, the Holweck Award of the French Physical Society and the Institute of Physics, and the Gold Medal of the C.N.R.S. Dr. Nozières' work has been concerned with various facets of the many-body problem, and his work currently focuses on crystal growth and surface physics.

David Pines
Center for Advanced Study Professor of Physics at the University of Illinois at Urbana-Champaign, he has made pioneering contributions to an understanding of many-body problems in condensed matter and nuclear physics, and to theoretical astrophysics. Editor of Addison-Wesley's *Frontiers in Physics* series and the American Physical Society's *Reviews of Modern Physics*, Dr. Pines is a member of the National Academy of Sciences and the American Philosophical Society, a Foreign Member of the Academy of Sciences of the U.S.S.R., and is a Fellow of the American Academy of Arts and Sciences and the American Association for the Advancement of Science. Dr. Pines received the Eugene Feenberg Memorial Medal for Contributions to Many-Body Theory in 1985, the P.A.M. Dirac Silver Medal for the Advancement of Theoretical Physics in 1984, and the Friemann Prize in Condensed Matter Physics in 1983.

Special Preface

We began writing this book at a time when field theoretical methods in statistical mechanics were expanding rapidly. Our aim was to focus on the physics which lies behind such sophisticated techniques, to describe simple physical facts in a simple language. Hence our deliberate choice of "elementary" methods in explaining such fundamental concepts as elementary excitations, their interactions and collisions, etc.... Rather than elaborating on calculations, we tried to explain *qualitative* and *unifying* aspects of an extremely broad and diversified field. Such a limited scope—albeit ambitious—probably explains why our book has retained popularity throughout the years. It is a comforting thought to evolve from a "frontier" level to a "classic" status. We hope it is not only a matter of age!

The book was originally organized in two volumes. Volume I dealt with "normal" Fermi fluids, *i.e.*, those which display no order of any type. Typical examples are liquid ^{3}He or electron liquids at temperatures above a possible superfluid transition. We discussed at length the nature of elementary excitations, the central concept of response functions, the new features brought about by the long range of Coulomb interactions in charged systems. Volume II was supposed to deal with superfluid systems, both bosons (^{4}He) and fermions (metallic superconductors); it was never completed. The main reason was a matter of timing. The year, 1965, marked an explosive growth of the work on superconductors, with such new concepts as phase coherence, the Josephson effect, etc. Things were moving fast, while our ambition was to provide a carefully thought out picture, in which concepts and methods were put in perspective. It was definitely not the appropriate time, and consequently Volume II was put aside. We nevertheless had completed a long chapter on Bose condensation and liquid ^{4}He, which has been widely circulated in the community. After some hesitation, we have decided to take the opportunity of this "classic" series to publish this chapter, written in 1964, as Volume II of our text. We do this partly because it contains physical concepts that have perhaps not been pursued in the detail they deserve (e.g., the interaction of elementary excitations), partly because we hope our early work will provide a perspective on the field of ^{4}He which will help the reader appreciate the subsequent evolution of ideas.

We decided to leave the text as it stands, despite some obvious weaknesses from a contemporary perspective. Had we begun to change it, we would have written it completely—a task which we could not complete now. It is clear today we would shorten many parts. We would often look at aspects in a different light, especially on the central concept of phase coherence. Nevertheless, the main issues we emphasized in 1964 are still valid—and not always fully appreciated.

Of course, research on liquid ^{4}He has not stopped during these past 25 years, and many new experimental and theoretical results have appeared on liquid ^{4}He. The excitation spectrum has been measured with extreme accuracy, including a detailed study of line width. The dispersion is now known to be anomalous, contrary to what we believed at that time. The roton interaction has been studied using Raman scattering: we now know that two rotons possess a bound state—and a "roton liquid" theory has been put forward, in the same spirit as the Landau theory of Fermi liquids. On a more macroscopic note, the physics of vortices has been pursued, but some of the main issues (nucleation of vortices, superfluid turbulence, etc...) have not yet been fully clarified. A recent spectacular development is the unambiguous observation by Varoquaux and Avnel of a Josephson effect in liquid ^{4}He, seen as produced by quantized phase dissipation in a Helmholtz resonator slippage. This experiment opens a new avenue of research for the future.

On the theoretical side, progress has been more limited. The Feynman program of calculating the phase transition has been completed by Ceperley with the aid of a Cray supercomputer, while good molecular dynamics simulations of the equation of state and variational and phenomenological calculations of the excitation spectrum which yield good agreement with experiment have been carried out.

However, much of the underlying physics is still not resolved. Is there some short range crystalline order in the liquid? Is that order responsible for the large depletion of the condensate population? Can there be a "normal Bose liquid," i.e., can depletion be complete? Of course, the question is irrelevant for ^{4}He, which is superfluid; nevertheless, it may be the key to understanding strong coupling situations. Semi-phenomenological approaches shed some light on such questions, but altogether there is no satisfactory theory of strongly coupled Bose liquids.

Shortly after the present text was written, two important monographs on ^{4}He appeared. One, *The Theory of Superfluidity*, by I. M. Khalatnikov, is entirely devoted to theory. It contains a detailed and authoritative survey of superfluid hydrodynamics, including a discussion of quasiparticle kinetics and dissipation, and has been reissued as a companion volume in this series. The second, *Liquid and Solid Helium*, by J. Wilks, remains, after some twenty years, the "bible" for all those working in that field. Only part of it deals with superfluid ^{4}He—but that part contains an extensive survey of the experimental situation as of 1967. The corresponding theory is developed in detail, with special emphasis on sound propagation and temperature processes. It is clear that our text overlaps that of Wilks in many places, but the emphasis is very different. We were mainly concerned with the conceptual background of superfluidity, viewed as a novel ordered state of condensed matter, while Wilks' theoretical developments are oriented towards experiments. Our text is thus in some way complementary to the Wilks monograph, biased toward statistical physics more than low temperature physics.

It had been our original intention to develop a comparable treatment for superconductors, and pursue the comparison between Bose liquids and superconductors (part of that was written in 1964—but we hardly remember what was supposed to come in chapters 8 to 11!). Our hope is that the text *as it is* may be of some help in clarifying the underlying physics of superfluid ^{4}He. In publishing it, we mostly respond to the demand of a number of colleagues, both old and young. We are quite aware of its "antique" character—but, although the presentation may at times be old-fashioned, and clearly reflects our youthful enthusiasm (writing a review is not such a bad way to learn a subject), we feel the physical picture is still correct. The alternative to not publishing would have been to let this text disappear in oblivion. We hope that the number of those who will find it useful will justify the publication delay of 25 years.

Philippe Nozières
David Pines

November, 1989

Preface

Our aim in writing this book has been to provide a unified, yet elementary, account of the theory of quantum liquids. Strictly speaking, a quantum liquid is a spatially homogeneous system of strongly interacting particles at temperatures sufficiently low that the effects of quantum statistics are important. In this category fall liquid ^{3}He and ^{4}He. In practice, the term is used more broadly, to include those aspects of the behavior of conduction electrons in metals and degenerate semiconductors which are not sensitive to the periodic nature of the ionic potential. The conduction electrons in a metal may thus be regarded as a normal Fermi liquid, or a superfluid Fermi liquid, depending on whether the metal in question is normal or superconducting.

While the theory of quantum liquids may be said to have had its origin some twenty-five years ago in the classic work of Landau on ^{4}He, it is only within the past decade that it has emerged as a well-defined subfield of physics. Thanks to the work of many people, we possess today a unified point of view and a language appropriate for the description of many-particle systems. We understand where elementary excitations afford an apt description and where they do not; we appreciate the relationship between quasiparticle excitations and collective modes, and how both derive from the basic interactions of the system particle. There now exists a number of model solutions for the many-body problem, solutions which can be shown to be valid for a given class of particle interactions and system densities: examples are an electron system at high densities and low temperatures, and a dilute boson system at low temperatures. In addition, there is a semiphenomenological theory, due to Landau, which describes the macroscopic behavior of an arbitrary normal Fermi liquid at low temperatures. Finally, and most important, has been the development of a successful microscopic theory of superconductivity by Bardeen, Cooper, and Schrieffer.

These developments have profoundly altered the main lines of research in quantum statistical mechanics: it has changed from the study of dilute, weakly interacting gases to an investigation of quantum liquids in which the interaction between particles plays an essential role. The resulting body of theory has developed to the point that it should be possible to present a coherent account of quantum liquids for the non-specialist, and such is our aim.

In writing this book, we have had three sorts of readers in mind:

(i) Students who have completed the equivalent of an undergraduate physics major, and have taken one year of a graduate course in quantum mechanics.

(ii) Experimental physicists working in the fields of low-temperature or solid-state physics.

(iii) Theoretical physicists or chemists who have not specialized in many-particle problems.

Our book is intended both as a text for a graduate course in quantum statistical mechanics or low temperature theory and as a monograph for reference and self-study. The reader may be surprised by its designation as a text for a course on statistical mechanics, since a perusal of the table of contents shows few topics that are presently included in most such courses. In fact, we believe it is time for extension of our teaching of statistical mechanics, to take into account all that we have learned in the past decade. We hope that this book may prove helpful in that regard and that it may also prove useful as a supplementary reference for an advanced course in solid-state physics.

We have attempted to introduce the essential physical concepts with a minimum of mathematical complexity; therefore, we have not made use of either Green's functions or Feynman diagrams. We hope that their absence is compensated for by our book being more accessible to the experimentalist and the nonspecialist. Accounts of field-theoretic methods in many-particle problems may be found in early books by the authors [D. Pines, *The Many-Body Problem*, Benjamin, New York (1962), P. Nozières, *The Theory of Interacting Fermi Systems*, Benjamin, New York (1963)] and in L. P. Kadanoff and G. Baym, *Quantum Statistical Mechanics*, Benjamin, New York (1962), and by A. A. Abrikosov, L. P. Gor'kov, and I. E. Dzyaloshinski, *Methods of Quantum Field Theory in Statistical Physics*, Prentice-Hall, New York (1964), to mention but a few reference works.

The decision to publish the book in two volumes stems, in part, from its length, and in part, from the natural division of quantum liquids into two classes, normal and superfluid. A third and perhaps controlling factor has been that a single volume would have meant a delay in publication of the present material of well over a year.

Although our book is a large one, we have not found it possible to describe all quantum liquids, or every aspect of the behavior of a given liquid. For example, we have not included a description of nuclear matter, of phase transitions, or of variational calculations of the ground state of various many-particle systems. On the other hand, we have compared theory with experiment in a number of places and, where appropriate, have compared and contrasted the behavior of different quantum liquids.

We have chosen to begin the book with the Landau theory of a neutral Fermi liquid in order to illustrate, in comparatively elementary fashion, the way both quantum statistics and particle interaction determine system behavior. We next consider the description of an arbitrary quantum liquid; we discuss the mathematical theory of linear response and correlations, which establishes the language appropriate for that description. We then go on to discuss, in Volume I, charged Fermi liquids, and in Volume II, the superfluid Bose liquid.

The authors began work on this book in Paris, at the Laboratoire de Physique of the Ecole Normale Superiéure, in the fall of 1962, when one of us was on leave from the University of Illinois. Since then we worked on the book both in Paris and at the University of Illinois. We should like to thank Professor Yves Rocard, of the Université de Paris, and Professor G. M. Almy, head of the Physics Department at the University of Illinois, for their support and encouragement. One of us (D.P.) would also like to thank the John Simon Guggenheim Memorial Foundation for their support during 1962 and 1963, and the Army Research Office (Durham) for their support during 1963 and 1964.

We should like to express our gratitude to the many friends and colleagues to whom we have turned for advice and discussion, and particularly to Professor John Bardeen for his advice and encouragement. We are deeply indebted to Dr. Conyers Herring for his careful reading of a preliminary version of Chapter 1, and to Professor Gordon Baym, who read carefully the entire manuscript and whose comments have improved both the accuracy and the clarity of our presentation. We owe an especial debt of gratitude to Dr. Odile Betbeder-Matibet, who has been of substantial assistance in the correction of the proof, and to Mme. M. Audouin, who has helped in the preparation of the index.

David Pines
Philippe Nozières

December 1966

Introduction

We begin our study of the superfluid Bose liquid with a brief review of the experimental and theoretical background of the liquid ^4He, the only Bose liquid found naturally in the laboratory. We then consider, in Chapter 2, the quasi-particle excitations and show how at $T = 0$ the macroscopic occupation of the zero momentum state leads, in the long wavelength limit, to an identical excitation spectrum for quasi-particles and density fluctuations. The spectrum consists in phonons with a velocity given by the macroscopic velocity of sound. We turn in Chapter 3 to a consideration of the observed elementary excitation spectrum of He II. We discuss both the experimental measurements and the theories that have thus far been proposed to explain them.

In Chapter 4 we begin our study of superfluid flow. We first consider the response of the superfluid system to a transverse probe (i.e., a rotating bucket). The transverse current response function provides a natural way to describe the response of the liquid to a rotation of its container. We show that for a Bose liquid the transverse static long wave-length response function vanishes at $T = 0$; this result guarantees that the superfluid will not follow a slow rotation of its container. We further discuss the long-range order which is characteristic of the superfluid state and introduce the important concept of a coherence length.

The results obtained in Chapters 2, 3, and 4 we may classify as characteristic of a "Bose liquid theory." They are based on the single important assumption that for a homogeneous Bose liquid at rest one has macroscopic occupation of a single quantum state, that of zero momentum. With that assumption, and a knowledge of the behavior of the various sum rules in the macroscopic (long wave-length) limit, one can derive both the excitation spectrum and the response functions in this limit. One thus succeeds with a minimum of calculation, and mathematical assumption, in deriving the important "superfluid" properties of an arbitrary Bose liquid. The theory in this way provides a direct link between macroscopic occupation of a single quantum state and the superfluidity of a Bose system.

In Chapter 5 we relax the restrictions that the condensate be at rest and uniform. We study superfluid flow, the quantization of circulation in the superfluid, and the elementary excitations in a moving superfluid. Superfluid behavior at finite temperatures is considered in Chapter 6. Particular attention is paid to the development of a microscopic basis for the two-fluid theory. A response-function definition of ρ_n is presented, and the Landau expression for that quantity is discussed in detail. In Chapter 7, the three kinds of sound wave propagation (first sound, second sound, and zero or quasi-particle sound) which are possible at finite temperatures are discussed. Particular attention is paid to the behavior of the density fluctuation excitation spectrum in different regimes of physical interest. In Chapter 8 we take up the question of vortex lines in the superfluid. We consider their dynamic properties and their mutual interaction; we then make use of these properties in an attempt to understand both the rotating bucket and critical current experiments with He II.

We next consider the microscopic calculation of the elementary excitations in interacting Bose systems. At present there exists a satisfactory microscopic theory for only two limiting cases:

1. A weakly interacting Bose gas [Bogoliubov (1947)]

2. A dilute Bose gas with an arbitrary law of interaction between the particles [Lee, Huang, and Yang (1957)].

Despite the fact that neither of the above cases corresponds to He II, such "model" calculations are extremely valuable. They provide explicit examples of the way one must build the macroscopic occupation of a single quantum state into the theory, and tell us, under well-defined circumstances, what the consequences of that occupation will be. In Chapter 9 we develop the microscopic theory of Bogoliubov et al. for the case of a uniform condensate, while in Chapter 10 we consider its generalization by Penrose and Onsager (1956), Gross (1961), and Pitaevskii (1961b) to the case of a non-uniform condensate. We conclude with a brief discussion of some of the open questions in the theoretical and experimental study of He II.

Contents

---

[1]The treatment in this section follows closely that in Hugenholtz and Pines (1959) and Pines (1963).

The Theory of Quantum Liquids

Superfluid Bose Liquids

CHAPTER 1

EXPERIMENTAL AND THEORETICAL BACKGROUND ON HE II

1.1 Introduction

We study a system of interacting bosons at very low temperatures. We assume that the density of the system is such that it remains liquid at absolute zero, a condition which is, in fact, satisfied only by ^{4}He. For helium, the forces are relatively weak, while, because of its low mass, the zero-point oscillations of the individual atoms are large. As a result, as London (1938) has shown, helium may be expected to remain a liquid down to the lowest temperatures. (Hydrogen solidifies because of the much stronger inter-atomic forces, while the heavier inert gases solidify at low temperatures because their zero-point oscillations possess lower amplitude.)

Liquid helium at temperatures below 2.2°K is not merely a quantum liquid in the sense that a quantum-mechanical description is essential for an understanding of its properties. It is, as well, a superfluid, which displays, amongst other properties, that of flowing through narrow channels with no measurable viscosity. Superfluidity is, as we shall see, to be expected of any Bose liquid at sufficiently low temperature. We shall see that a certain number of rigorous results may be derived for such a superfluid system; we may call the corresponding body of theory, "Bose liquid theory."

Because there exists in nature only one Bose liquid, ^{4}He, it would seem appropriate that we begin with an account of some of the essential experimental facts concerning this remarkable liquid. To a considerable extent, the unravelling of the properties of ^{4}He has proceeded from a close collaboration between theorists and experimentalists. We shall therefore attempt to recount as well some of the essential steps in the development of our theoretical understanding of liquid ^{4}He. Our historical account of both experiment and theory will be brief; the interested reader is referred to the books of London, Atkins, and Lane for a more detailed exposition.

1.2 Early Experimental Evidence for the Superfluid Phase

The critical temperature for the liquefaction of ^{4}He is 5.2°K. At 2.19°K the liquid undergoes a phase transition characterized by no latent heat, and a specific heat behavior shown in Fig. 1.1. The transition is often referred to as a "λ" transition, because of the shape of the specific heat curve. The low temperature phase is called He II; the high temperature phase, He I. Indications of the existence of He II were already present in

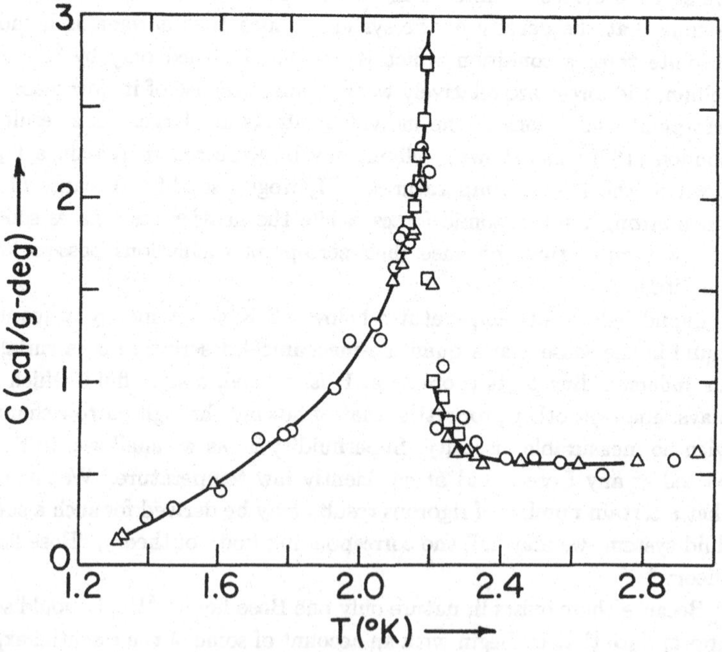

FIGURE 1.1. *Specific heat of liquid helium II [after Keesom and Clausius (1932)].*

the early experiments of Kamerlingh Onnes, who found in 1911 that the liquid when cooled below 2.2°K began to expand rather that to contract. In 1924, Kamerlingh Onnes and Boks showed the density as a function of temperature possessed a sharp maximum at 2.19°K; in 1932 the nature of the "λ" transition was clarified by the specific heat measurements of Keesom and Clausius shown in Fig. 1.1. However, it was not until some six years later that low temperature physicists, working in Cambridge and Moscow, discovered that liquid He II was in fact a quite new kind of liquid, which might be appropriately called a "superfluid" to distinguish its behavior from that of other "normal" fluids.

The "superfluidity" of He II was discovered independently, and essentially, simultaneously, by Kapitza (1938) and Allen and Misener (1938), by studying the flow of liquid helium through a thin capillary. They found that as the liquid was cooled below the λ-point (the temperature at which the phase transition takes place) the viscosity dropped by many orders of magnitude to an essentially unmeasurable value.

Shortly thereafter Allen and Jones (1938) discovered the "fountain effect," a striking manifestation of the superfluid properties of He II. A simplified version of their experimental apparatus is shown in Fig. 1.2a;

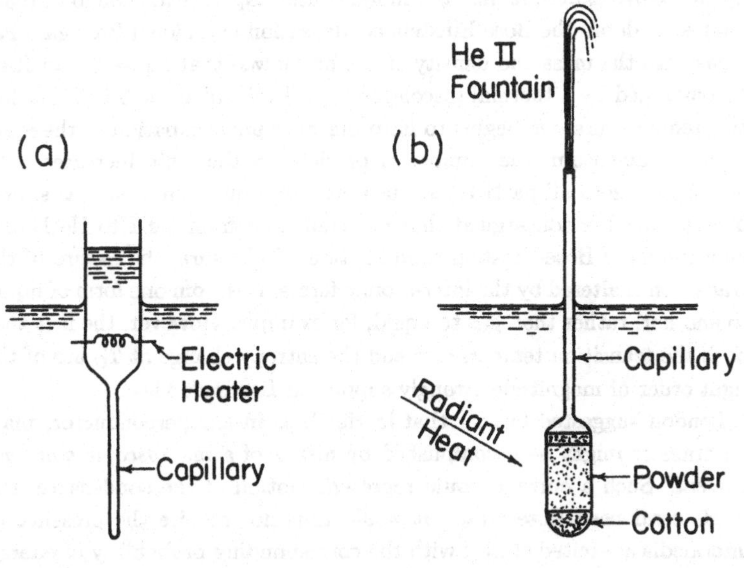

FIGURE 1.2. *Schematic representation of fountain effect experiments of Allen and Jones (1938) [after Atkins (1959)].*

there one sees two reservoirs, A and B, connected by a thin capillary. Allen and Jones found that when heat was supplied to the inner part (A), the level of the liquid rose. A more dramatic version of their experiment is shown in 1.2b; when the emery powder at the bottom of the tube is heated, a jet of liquid helium spurts out. Both versions of the experiment tell us that heat transfer is accompanied by matter transfer, a fact quite inexplicable according to the usual laws of hydrodynamics and thermodynamics.

That these laws do not apply to He II also followed from earlier experiments by Allen, Peierls, and Udwin (1937), who found that in flow through thin capillaries, the heat current was not proportional to the temperature gradient. The three experiments, taken together, showed that He II was a basically new kind of liquid, for which none of the familiar differential equations of thermodynamics and hydrodynamics governing heat and matter transfer were applicable.

1.3 Development of the Two-Fluid Model

Theoretical physicists were not long in responding to the challenge posed by the above experiments. In a fundamental paper, Fritz London (1938) studied in detail the Bose-Einstein condensation of an ideal Bose gas. For a gas with the mass and density of He, he showed that a phase transition accompanied by a thermal discontinuity will take place at 3.13°K. Below this temperature one begins to have *macroscopic* occupation of the state of zero-momentum; the number of particles in the state increases until at absolute zero all particles are in that single quantum state, as shown in Fig. 1.3. London argued that the transition from He I to He II was an example of Bose-Einstein condensation. To be sure, the nature of the transition is altered by the inter-atomic forces; it is from one form of liquid to another, rather than gas to liquid, for example. However, the fact that both the transition temperature and the entropy change at T_C are of the right order of magnitude strongly supported London's view.

London suggested further that in He II, as in a superconductor, matter transfer might be accomplished by means of a *macroscopic quantum current*. Such a current would represent motion of the condensate, the single condensed Bose state. It would thus not require the presence of intermediate excited states, with the corresponding probability of energy dissipation. Rather it would correspond to an adiabatic transformation of the condensate brought about by changes in the macroscopic boundary

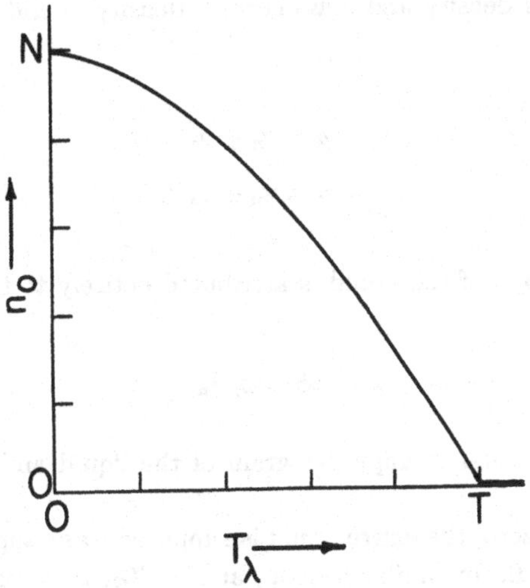

FIGURE 1.3. *Condensation of a free Bose gas [London (1938)].*

conditions. Thus it was London's view that "an understanding of a great number of the most striking peculiarities of liquid helium can be achieved without entering into any discussion of details of molecular mechanics, merely on the hypothesis that some of the general features of the degenerate ideal Bose-Einstein gas remain intact, at least qualitatively, for this liquid." We shall see that this view has turned out to be a correct one.

Tisza (1938, 1940, 1947) used this idea of London's as a basis for the development of a phenomenological *two-fluid model* for liquid He II. His theory succeeded not only in explaining the above experimental puzzles, but enabled him to predict new physical phenomena which were subsequently discovered experimentally. We sketch here briefly the basic ideas of the two-fluid model, as presently applied to He II.[1]

In the two-fluid model, superfluid He II is regarded as being composed of two fluids which are simultaneously present and mutually interpenetrating. The superfluid component is the *condensate*, the single condensed quantum state of the Bose liquid. Being a single quantum state, it carries no entropy. It is determined by the macroscopic boundary conditions, and may in general be characterized by a mass density, ρ_s and a velocity field $\mathbf{v}_s$. The second component, the normal fluid, of density ρ_n, velocity $\mathbf{v}_n$, is assumed to carry the entire thermal excitation of the liquid.

[1]Tisza's original theory was presented in somewhat different form from that given here.

The total mass density and mass current density, ρ and $\mathbf{J}$, are assumed to be given by:

$$\rho = \rho_n + \rho_s \tag{1.1}$$

$$\mathbf{J} = \rho_n \mathbf{v}_n + \rho_s \mathbf{v}_s \tag{1.2}$$

while the entropy of the liquid is attributed entirely to the normal fluid part, so that

$$\rho S = \rho_n S_n \tag{1.3}$$

where S is the total entropy per gram of the liquid and S_n the normal fluid entropy.

At absolute zero, the entire liquid is supposed to be superfluid, so that $\rho_s = \rho$ at $T = 0$. In similar fashion, at $T = T_C$, $\rho_n = \rho$: the superfluid component has vanished. Finally, there is no viscosity associated with superfluid flow (the condensate being a single quantum state), while the coefficient of viscosity, η_n, of the normal fluid is expected to be comparable to that of He I, and to go over to it at the λ-point.

The two-fluid theory offers a simple explanation of the two basic experiments we have described above. The existence of superfluid flow through a narrow capillary is, of course, built into the theory. Moreover the theory resolves an apparent paradox: that measurements of the viscosity of He II by means of a rotating disc method (Keesom and MacWood, 1937) had shown a finite value for the viscosity coefficient, which went over to that of He I at the λ-point. The paradox disappears once one realizes that a disc rotating in helium does not interact with the superfluid; it is, however, scattered by the normal fluid, and thus the rotation of a disc makes possible a determination of ρ_n and η_n.

The explanation of the fountain effect follows directly from the remark that according to the theory a thin capillary acts like an "entropy filter." The normal fluid, which carries all the entropy, will be scattered by the walls and cannot pass through. Only the superfluid passes through. Superfluid flow being a reversible process, one can apply thermodynamics to the problem of the two reservoirs connected by an "entropy filter." One finds directly that in equilibrium there will be a pressure difference associated with any temperature difference between the two containers; the higher pressure is associated with the higher temperature, in accordance with the experimental results.

On the basis of the above considerations, Tisza was led to purpose the mechanocaloric effect, which is essentially the inverse of the fountain effect. If the two containers, A and B, are to be kept under isothermal conditions, ($T_B < T_A$), then heat must be supplied to B and carried away from A. The effect was subsequently discovered by Daunt and Mendelssohn (1939). It received quantitative verification in a beautiful experiment by Kapitza (1941) who studied the reversible flow of helium between two optically polished quartz discs.

Perhaps the most striking success of the two fluid theory was the prediction of a new type of wave propagation in He II. Tisza developed a generalization of the usual hydrodynamic equations to describe the departure of the fluid densities, ρ_s and ρ_n, from their equilibrium values. He found that in addition to the usual hydrodynamic "first" sound wave, there should exist in He II a new wave, which he called second sound. First sound corresponds to a propagation of a local density fluctuation in which the two fluids oscillate in phase. Second sound represents a *temperature* wave (one involving propagation of a local fluctuation in temperature) in which the two fluids oscillate out of phase. Following an independent prediction by Landau of its existence, Shalnikov and Sokolov attempted, without success, to find second sound using a conventional quartz transducer to excite the wave. Subsequently, Lifshitz (1944) pointed out that a density wave is only weakly coupled to the second sound mode, so that their failure was not surprising. Soon thereafter Peshkov (1944, 1946) used a localized source of heat to excite the wave, and confirmed its existence.

1.4 Elementary Excitations: Theory and Experiment

An important theoretical development came with a remarkable paper by Landau (1941), in which the concept of elementary excitations was introduced to describe the low-lying excited states of He II. Landau concluded from quite general arguments that in a quantum liquid two kinds of excited states might occur. The first consists in potential flow and represents sound waves of velocity s. The corresponding elementary excitations are the usual phonons, with an energy vs. momentum relation,

$$\varepsilon_{\text{phonon}} = sp \tag{1.4}$$

The second sort of excited state would represent the creation of vortices in the liquid. Landau argued that such states require a finite amount of energy to excite; that the corresponding elementary excitation, which he called a "roton," would thus possess an energy-momentum relation,

$$\varepsilon_{\text{roton}} = \Delta + \frac{p^2}{2\mu} \tag{1.5}$$

Landau proposed that the normal fluid (in the two-fluid model) be regarded as a dilute gas of weakly-interacting thermally-excited elementary excitations. Elementary statistical considerations then permit explicit calculation of the specific heat, entropy, etc., associated with the normal fluid. The background liquid through which the excitations move corresponds to the superfluid component; it carries the remaining mass and momentum of the liquid. In some respects, Landau's work resembled that of Tisza; he defined v_n and v_s in essentially the same way, and was led to results similar to those of Tisza for the various macroscopic relations obtained within the two-fluid model. (These included an independent prediction of second sound; because of war-time conditions, Landau did not see Tisza's paper until after publication of his own.) Landau's paper represented, however, a most important step beyond the work of Tisza, because of his explicit identification of the two fluids in question. He was able thereby to relate the various macroscopic two-fluid parameters to the microscopic spectrum of elementary excitations of the liquid. He could thus derive an explicit expression for ρ_n and predict that at sufficiently low temperatures the velocity of second sound, s_2, would be related to that of first sound, s_1, according to

$$s_2^2 = s_1^2/3 \tag{1.6}$$

The result, (1.6), was subsequently verified by de Klerk, Hudson, and Pellam (1954).

The work of Landau stimulated a number of experiments, of which we mention two. The first is a direct measurement of ρ_n by Andronikashvili (1946). Landau had considered the rotation of a vessel of He II, and had argued that only the normal part of the liquid would be dragged along by the rotation of the vessel. Hence a measurement of its moment of inertia would furnish a direct measurement of ρ_n. Andronikashvili studied the torsional oscillations in He II of a pile of thin discs spaced closely together and suspended by a thin wire. The discs act to entrap only the normal fluid; a measurement of the frequency of oscillation furnishes a direct measure of the moment of inertia of the discs plus the entrapped helium, and hence of ρ_n.

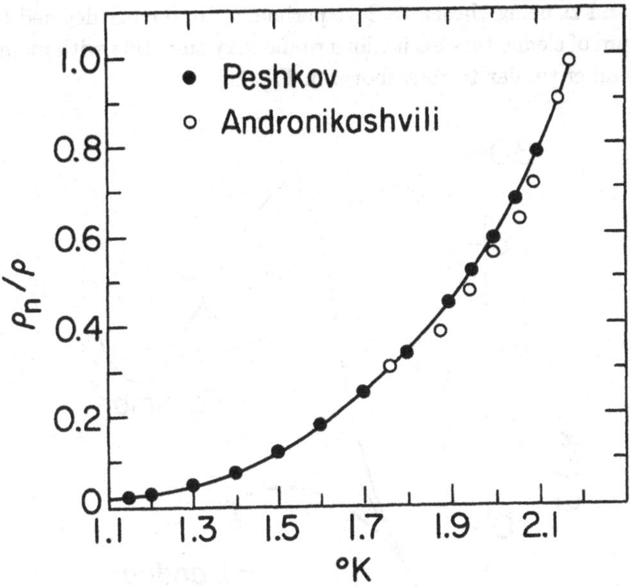

FIGURE 1.4. *Experimental measurements of ρ_n/ρ [from Peshkov (1946)].*

An independent measurement of ρ_n was obtained by Peshkov (1944, 1946) in his experiments on the velocity of second sound. According to the two-fluid model, the speed of propagation of the thermal wave is given by:

$$s_2^2 = \left(\frac{\rho_s}{\rho_n}\right)\frac{S^2T}{C_P} \tag{1.7}$$

where S is the entropy and C_P is the specific heat at constant pressure. A measurement of s_2 when combined with an independent measurement of the entropy thus enables one to determine ρ_n. The values obtained by Peshkov were in excellent agreement with those of Andronikashvili; their results are shown in Fig. 1.4. Moreover, Peshkov's experiment provided Landau (1947) with the data needed to work backwards from his theoretical expression for ρ_n to a form for the elementary excitation spectrum which might be consistent with experiment. In this way he was led to revise his proposed roton spectrum. Landau concluded that the rotons would possess a finite momentum, p_o, and that in the vicinity of p_o they would possess an energy vs. momentum relation,

$$\varepsilon_{\text{roton}}(p) = \Delta + \frac{(p-p_o)^2}{2\mu} \tag{1.8}$$

p_0, μ, and Δ being chosen to fit experiment. He further decided that the spectrum of elementary excitations would vary smoothly with momentum, in a manner similar to that shown in Fig. 1.5.

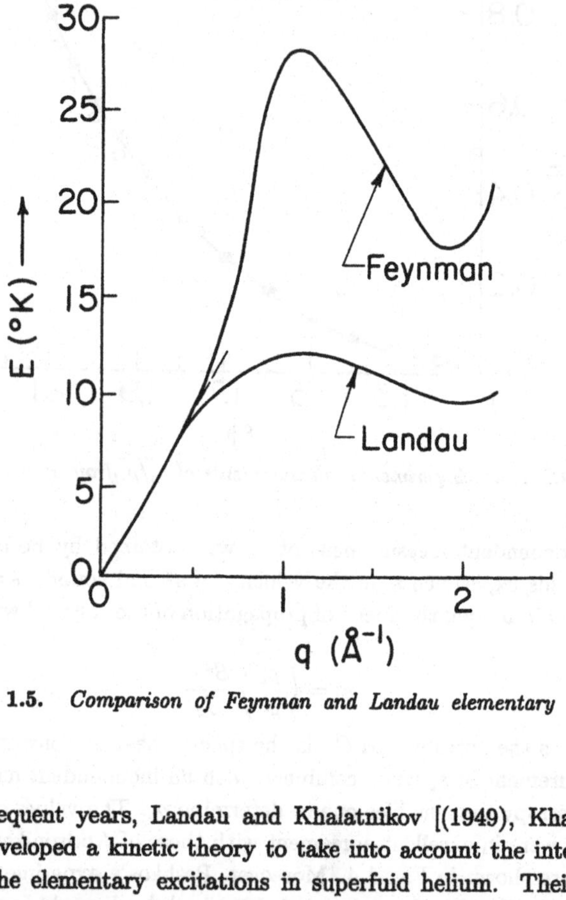

FIGURE 1.5. *Comparison of Feynman and Landau elementary excitation spectra.*

In subsequent years, Landau and Khalatnikov [(1949), Khalatnikov (1952)] developed a kinetic theory to take into account the interactions between the elementary excitations in superfuid helium. Their theory has made possible an understanding of many temperature-dependent irreversible phenomena, such as viscosity and the attenuation of second sound.

As we have mentioned, Landau was led to his proposed elementary excitation spectrum by considerations of quantum hydrodynamics; nowhere in his theory did the possibility of Bose-Einstein condensation enter explicitly. (Indeed, Landau rejected rather summarily the analogy between He II and the ideal Bose gas which had played such an important role in the London-Tisza theory.) A strong hint that Bose statistics were indeed of decisive importance came from the work of Bogoliubov (1947) who developed the first microscopic theory of an interacting boson system. Bogoliubov showed that as a consequence of macroscopic occupation of the zero-momentum state the long wave-length elementary excitation spectrum of weakly-interacting bosons is altered. In place of the familiar quadratic dependence of energy on momentum, characteristic of the free boson system, one finds a linear energy vs. momentum relation, such as (1.4). Subsequently Feynman (1955) stressed anew the great importance of Bose statistics in determining the elementary excitation spectrum of He II. He presented a series of elegant physical arguments concerning the nature of the low-lying excited states in a Bose liquid, from which he concluded that an elementary excitation of momentum p would possess an energy

$$\varepsilon_p^F = \frac{p^2}{2mS_p} \tag{1.9}$$

where S_p is the static form factor for the liquid. The latter may be obtained by x-ray or neutron scattering; the resulting excitation spectrum is likewise shown in Fig. 1.5. It may be seen that there is good qualitative agreement between the calculated Feynman spectrum and that inferred by Landau from measurements of ρ_n.

On the basis of this agreement one might conclude:

1. That the Bose statistics play an essential role in determining the excitation spectrum.
2. That rotons and phonons are part of the same spectrum.

The first of these conclusions is, of course, much buttressed by the vastly different experimental behavior of ^{3}He and ^{4}He. The second received further theoretical support from a subsequent calculation by Feynman and Cohen (1956) of the excitation spectrum of He II, one which agreed much better in the "roton" region with the Landau result. It received direct experimental confirmation from the experiments of Palevsky, Otnes and Larsson (1958), in which inelastic neutron scattering was used to measure the elementary excitation spectrum directly. An experimental result for the excitation spectrum, together with the theoretical prediction of Feynman and Cohen, is shown in Fig. 1.6. There one sees that the excitation spectrum does indeed pass smoothly from the phonon part through, and beyond, the roton region.

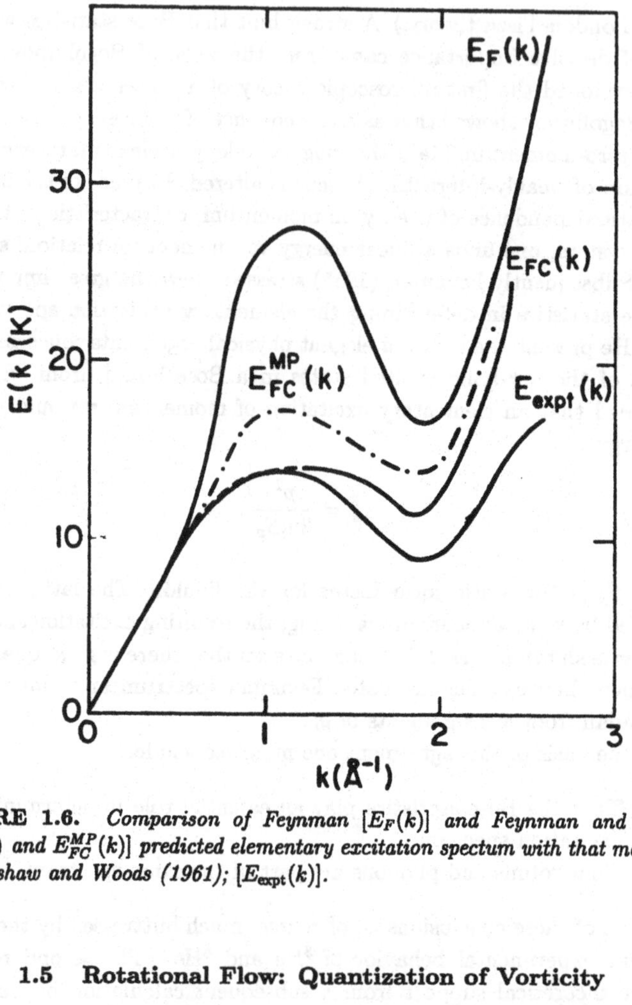

FIGURE 1.6. *Comparison of Feynman [$E_F(k)$] and Feynman and Cohen [$E_{FC}(k)$ and $E_{FC}^{MP}(k)$] predicted elementary excitation spectum with that measured by Henshaw and Woods (1961); [$E_{expt}(k)$].*

1.5 Rotational Flow: Quantization of Vorticity

Perhaps the most striking defect of the theory we have thus far sketched concerns the criterion for the existence of superfluid flow through a pipe. According to Landau, friction between the superfluid and the walls begins when the walls are able to create elementary excitations in the fluid flowing past them. Such creation cannot take place as long as the superfluid velocity, $\mathbf{v}_s$, satisfies the condition

$$\mathbf{v}_s \leq \varepsilon_p / p \tag{1.10}$$

where ε_p is the energy of an elementary excitation of momentum **p**. For He II, the condition (1.10) predicts a critical velocity for superfluid flow which is of the order of 65 meters/sec. Observed critical velocities depend somewhat on the size of the capillary, but are some two orders of magnitude smaller than this. It would thus seem that another kind of excitation is playing a role in limiting the superfluid flow velocity.

An important clue to the nature of that excited state was provided by Onsager (1949) who suggested that *quantized vortex lines* might exist in He II. The flow we have considered thus far is irrotational flow—and such flow is the only kind possible in a singly connected region. However, as is well known in classical hydrodynamics, one may obtain rotational flow by having vortex lines, that is, singularities around which the liquid is rotating. Onsager proposed that such lines exist in He II, and argued that the circulation around any closed circuit in the liquid would be quantized in units of h/m,

$$\oint \mathbf{v}_s \cdot d\ell = nh/m \tag{1.11}$$

where n is an integer. This quantization of circulation is, as we shall see, a direct and natural consequence of a description of the superfluid component as a single macroscopic quantum state.

Quantization of vorticity was proposed independently by Feynman (1955) who developed its consequences in some detail. Feynman showed that it was quite likely that the resistance to flow above a critical velocity was caused by the production of vortex lines through friction between the liquid and the walls of the vessel. He estimated the critical velocity for the production of vortices to be of the order of

$$v_o = \frac{\hbar}{md} \ell_n \frac{d}{a} \tag{1.12}$$

where d is the radius of the capillary and a the size of the vortex core. For a capillary of diameter 10^{-5} cm, with $a = 10^{-8}$ cm, the above critical velocity is of the order of 100 cm/sec, in good qualitative accord with experimental observations.

The existence of vortex lines in He II was established experimentally by Hall and Vinen (1956), who studied the propagation of second sound in a rotating vessel containing superfluid helium. As a result of the rotation, there are present vortex lines parallel to the axis of rotation. Hall and Vinen measured the attenuation of second sound propagating both parallel and at right angles to that axis. They found a much greater attenuation of the second sound propagating at right angles and attributed this additional attenuation to the scattering of the normal fluid (here rotons) by the vortex lines. Such vortex lines play a decisive role in determining a number of phenomena that take place once the critical velocity of superfluid flow is reached. One finds, for example, a *mutual friction* between the normal and superfluid components, a phenomena which finds its explanation in the scattering of phonons and rotons by the newly produced vortex lines.

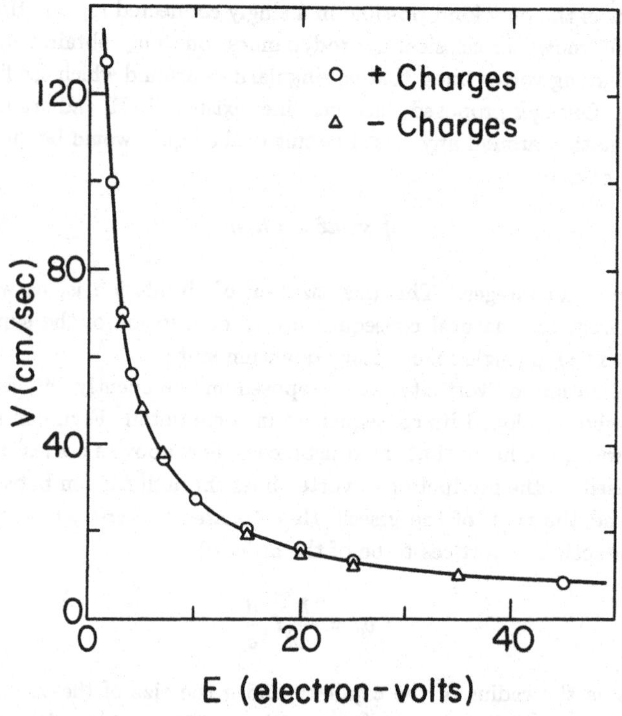

FIGURE 1.7. *Theoretical curve of velocity v vs. energy E of a vortex ring having one quantum of circulation* $k = h/m$ *and a core radius* $a = 1.2 \times 10^{-8}$ *cm [from Rayfield and Reif (1963)].*

Subsequently Vinen (1958, 1961a,b) was able to detect a single quantized vortex line, a remarkable achievement in view of the very small energies involved. (For liquid helium at $T = 0$ a vortex line with a single quantum of circulation, in a bucket of reasonable size, possesses a kinetic energy of the order of 10^{-7} erg/cm.)

Rayfield and Reif (1963) have discovered the existence of single quantized charged vortex rings in superfluid helium. In studying the motion of both positive and negative ions in He II at temperatures between 0.3°K and 0.6°K, they found that the "ions" moved as free particles, which could be accelerated or decelerated at will by the appropriate combination of applied electric fields. Quite surprisingly, when the energy of the "ions" increased, their velocity decreased. Such behavior suggested strongly that the ions were trapped in a vortex ring, and that it was the motion of such rings which was under observation. In Fig. 1.7 we reproduce a comparison between the classical hydrodynamic calculation of the energy vs. velocity curve for a vortex ring, and the experimental results of Rayfield and Reif. The agreement between theory and experiment leaves no doubt that they have observed the motion of a single vortex ring, which possesses one quantum of circulation, and has trapped within it a charged "ion."

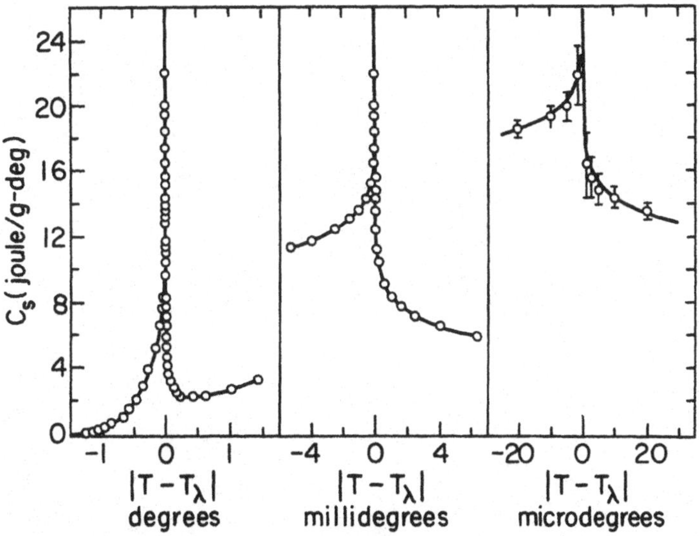

FIGURE 1.8. *Specific heat of liquid helium vs.* $T - T_\lambda$ *[from Buckingham and Fairbank (1961)].*

1.6 Nature of the λ-Transition

We conclude this short survey of the experimental data with a brief discussion of the elegant experimental work of Buckingham and Fairbank (1961) on the nature of the λ-transition in He II. They have measured the specific heat of He II to within 10^{-6} degree of the λ-point, with equal precision both above and below T_c. Their results are shown in Fig. 1.8; they show clearly the existence of a logarithmic singularity in the specific heat at $T = T_c$. Indeed, Fairbank and Buckingham find that over a factor of 10^4 in $(T - T_c)$ their data can be fit with an expression of the following form:

$$C_s = 4.55 - 3.00 \log_{10} |T - T_c| - 5.20\Delta$$

where $\Delta = 0$ for $T < T_c$ and $\Delta = 1$ for $T > T_c$. No satisfactory theory of this logarithmic singularity has as yet been proposed; explaining the details of the phase transition in He II remains a major theoretical problem.

CHAPTER 2

ELEMENTARY EXCITATIONS

2.1 Non-Interacting Bose Gas

The eigenstates of a gas of non-interacting bosons are symmetrized combinations of N different single particle states. As for the case of the corresponding fermion system, the eigenstates may be characterized by a distribution function, n_p, equal to 1 if the state p is occupied, zero otherwise. The system energy is

$$E = \sum_p n_p \frac{p^2}{2m} \tag{2.1}$$

The ground state is obtained by putting all the particles in the state $p = 0$ (there being no restriction on the number of particles in a single quantum state). The number of particles in this state is

$$N_o = N \gg 1 \tag{2.2}$$

One has, therefore, *macroscopic occupation of a single quantum state.* It is this feature, which persists in the interacting Bose liquid, [although not in the form (2.2)], which provides the essential clue to an understanding of the superfluid behavior of Bose liquids.

Consider now a state with one added particle of momentum $\mathbf{p}$. The energy of that state will be simply

$$\varepsilon_p^o = \frac{p^2}{2m} \tag{2.3}$$

Suppose, on the other hand, that one creates a density fluctuation of momentum $\mathbf{q}$. For the non-interacting boson system, this may only be done by taking a particle out of the zero-momentum condensate, and giving it a momentum **q**. The energy of the density fluctuation is

$$\omega_q^o = \frac{q^2}{2m} \tag{2.4}$$

We see that for the same momentum the particle spectrum and the density fluctuation spectrum are identical.

This feature of system behavior is quite different from that found for the fermion system. For a free fermion gas, as we have seen in Chapter 2, Vol. I, one may create a density fluctuation of momentum **q** by exciting a particle-hole pair of energy $\varepsilon^o_{\mathbf{p}+\mathbf{q}} - \varepsilon^o_{\mathbf{p}}$. The density fluctuation excitation spectrum is continuous, since all particle transitions allowed by the Pauli principle will contribute. The quite different behavior of the boson gas is a direct consequence of the fact that all particles are in the zero momentum state; there is thus only *one* way to create a density fluctuation of momentum **q**. As a result, the free boson density fluctuation spectrum resembles more nearly that found for collective modes in a fermion system than that found for free particles.

2.2 Quasi-particles

Let us apply to an interacting Bose liquid the approach developed in Chapter 1, Vol. I, for the interacting Fermi liquid, that of adiabatically turning on the interaction between the particles. We may expect to generate in this fashion the ground state of the real, interacting system from the ground state of the non-interacting Bose gas. As the interaction is turned on, the original ground state ($N_o = N$), is coupled to configurations in which some of the condensed particles are virtually excited into states with non-zero momentum. The particle interaction thus acts to deplete the condensed state $p = 0$. We shall assume that this depletion is not complete, so that the macroscopic occupation of the state of zero momentum persists in the Bose liquid. In the presence of particle interaction, the number of particles in that state, N_o, will no longer be equal to N, but it will still be a macroscopic number, such that

$$N_o \gg 1 \tag{2.5}$$

This assumption may be thought of as characteristic of a "well-behaved" Bose liquid, in the same sense that the existence of a well-defined Fermi surface, across which ε_p is continuous, characterizes a "normal" Fermi liquid.

Let us now add a particle with momentum **p** to the ground state of the ideal Bose gas, and again turn on the interaction between the particles adiabatically. In this fashion we generate an excited state of the Bose liquid which corresponds to its real ground state plus a *quasi-particle* of momentum **p**. If the added particle has momentum zero, we generate thereby the ground state of the $N+1$ particle system, which possesses an energy

$$E_o + \mu$$

where μ is the chemical potential, and E_o is the ground state energy for the N particle system. If the added particle has momentum **p**, we generate a state of energy

$$E_o + \mu + \varepsilon_p$$

where ε_p is the free energy of the quasi-particle, that is, the energy of the quasi-particle measured with reference to the chemical potential.

We can view the quasi-particle in much the same way as we did for the Fermi liquid, as a bare particle moving together with the surrounding distortion brought about by particle interactions. There is, however, an important new feature characteristic of the Bose liquid: a state with one extra quasi-particle does *not* contain one extra bare excited particle. To see this, we remark that as a result of the coupling between the added particle and the particles in the condensate (the zero momentum state), part of the added particle may be absorbed in the condensed phase (or vice-versa). This would cause no complications if one were able to keep careful track of the number of particles in the condensed phase. *However*, and this is an essential point, it is neither convenient, nor in fact, necessary, to do this. Because one has macroscopic occupation of the condensed state, one can (and does) neglect small fluctuations in the number of particles, N_o, in that state. (The approximation is valid up to terms of order of $1\sqrt{N}$.) Once one has adopted this point of view, the number of quasi-particles is clearly no longer conserved.

We may expect that a state with one quasi-particle is well-defined only to the extent that there are few states available into which the quasi-particle can decay. Again, only if the quasi-particle lifetime is long, will the adiabatic switching process we have employed be physically meaningful. Let us consider the simplest possible scattering process, one in which a quasi-particle of momentum **p** scatters against a particle in the condensate into two new quasi-particle states $\mathbf{p}'$ and $\mathbf{p}-\mathbf{p}'$. The scattering process will be real, and limit the lifetime of the quasi-particle $\mathbf{p}$, provided energy is conserved in the scattering process, a condition which reads:

$$\varepsilon_{\mathbf{p}} = \varepsilon_{\mathbf{p}-\mathbf{p}'} + \varepsilon_{\mathbf{p}'} \tag{2.6}$$

Since we do not keep track of the number of particles in the condensate, the process appears as the decay of the quasi-particle into two new quasi-particles, and represents another example of the lack of quasi-particle conservation.

The importance, or indeed the existence, of the mechanism (2.6), depends on the quasi-particle energies and the phase space available for the decay process. We shall see that for liquid He II, the criterion (2.6) is only met for $\mathbf{p}$ greater than a critical momentum $\mathbf{q}_c$, and that once decay of one quasi-particle into two becomes possible it goes quite rapidly. Effectively, $\mathbf{q}_c$ represents a cut-off momentum beyond which a quasi-particle can no longer be regarded as well-defined.

Higher order collision processes can also limit the lifetime of a quasi-particle. Thus a quasi-particle of momentum $\mathbf{p}$ can decay into three other quasi-particles (whose total momentum is $\mathbf{p}$), four, etc. In general we may expect that the amount of phase space available for the higher order processes will reduce their effectiveness; certainly where (2.6) is satisfied, decay into two quasi-particles is the most important decay mechanism.

We see that matters are somewhat less precise for a Bose liquid than its Fermi counterpart. For an arbitrary particle interaction, there is no *a priori* region of momenta for which quasi-particles exist as well-defined elementary excitations of the system. In fact, matters are not quite this bad; as we shall see, the long wave-length quasi-particle excitations turn out to be phonons, with a linear dispersion law. If the linear dispersion law were rigorous, a process like (2.6) would make only a negligible contribution to the phonon lifetime, since it would occur only for the special case that $\mathbf{p}, \mathbf{p}'$ and $\mathbf{p}-\mathbf{p}'$ lie along the same line. The importance of (2.6) depends then on departures from linearity of the phonon dispersion law. If the energy vs. momentum curve for the phonons possesses downward curvature, i.e.,

$$\varepsilon_p = sp + bp^3 (b < 0) \tag{2.7}$$

then the decay process, (2.6), is impossible. If, however, the curvature is upward ($b > 0$), then two-phonon decay can occur. Depending as it does on curvature for existence, one expects that probability of phonon decay via (2.6) to be proportional to some power of the phonon momentum higher than the first. As a result sufficiently long wave-length phonons represent well-defined excitations of the Bose liquid. We shall see, on the basis of sum rule arguments in the following section, that this is indeed the case.

2.3 Equilibrium Distribution of Quasi-Particles at Finite Temperatures

At finite temperatures the free Bose gas is characterized by a distribution function:

$$n_p^o(T) = \frac{1}{\exp[\beta\varepsilon_p^o] - 1} \tag{2.8}$$

$$n_o^o = N - \sum_p n_p^o(T) \tag{2.9}$$

This situation persists up to the transition temperature, T_c, at which one has:

$$\sum_p n_p^o(T_c) = N \tag{2.10}$$

$$n_o^o = 0 \tag{2.11}$$

The transition temperature is thus characterized by a transition from macroscopic occupation of the zero-momentum state, with n_o comparable to N, to microscopic occupation of that state, with $n_o \sim 1$.

To what extent may we hope that similar considerations apply in the case of the Bose liquid? First, by analogy with the case of the Fermi liquid, we might expect that the quasi-particles, where well-defined, may be described by a distribution function n_p which has exactly the same form as that pertaining to the free Bose gas, but in which the quasi-particle energy will appear. Thus one is led to the finite temperature distribution function proposed by Landau (1941):

$$n_p(T) = \frac{1}{\exp[\beta\varepsilon_p] - 1} \tag{2.12}$$

$$n_o(T) = N - \sum_p n_p(T) \tag{2.13}$$

At $T = 0$, the quasi-particle distribution function reduces to

$$n_o(o) = N \tag{2.14}$$

$$n_p(o) = 0 \tag{2.15}$$

It is to be expected that a relation like (2.12) is only approximate, since it implies that the excitations may be treated as a dilute, non-interacting *gas*. More generally we would expect to find an excitation *liquid*, in which, as for the Fermi liquid, the quasi-particle energy, ε_p, depends on the distribution function, n'_p, of all the other quasi-particles. The corresponding correction to ε_p may be viewed as an interaction energy between quasi-particles. A microscopic theory which incorporates this feature has been developed by Balian and de Dominicis (1965); it is, unfortunately, much more complicated that the corresponding Fermi liquid theory, for reasons we shall discuss later. Consequently, we do not attempt to present a specific Bose liquid theory, but confine ourselves to some general remarks.

We first note that as a consequence of quasi-particle interaction, the energy of a given quasi-particle, ε_p, will be a function of T. It then seems reasonable to write the equilibrium distribution function as

$$n_p(T) = \frac{1}{\exp[\beta\varepsilon_p(T)] - 1} \tag{2.16}$$

which is the form found by Balian and de Dominicis. The corresponding temperature variation of n_p and ε_p is by no means simple. The complexity of the problem stems from the fact that we are interested in temperatures of the order of T_c, for which the thermal corrections to the quasi-particle energy ($\sim kT/\mu$ or T/T_c) are important. Such complications did not appear in the greater part of our study of the Fermi liquid, since we confined our attention to the extreme degenerate domain, where $kT \ll \mu$ and thermal corrections are negligible. On the other hand, for ^{3}He at temperatures close to the degeneracy temperature (μ/k), a theory of comparable complexity is required. As long as we confine our attention to temperatures $T \ll T_c$, the number of excited quasi-particles is small; we expect the quasi-particle energy ε_p to be very close to its zero-temperature value. The simple result, (2.12), is then valid in the same sense as is the Landau Fermi liquid theory presented in Chapter 1, Vol. I.

Let us emphasize that n_p represents the distribution of quasi-particles. It is not to be identified with the quantity

$$N_p = \langle a_{\mathbf{p}}^{+} a_{\mathbf{p}} \rangle \tag{2.17}$$

which represents the *bare* particle distribution, averaged over the statistical distribution appropriate to the interacting Bose liquid. Consider for instance

$$N_o = \langle 0|a_o^+ a_o\,|0\rangle \tag{2.18}$$

where $|0\rangle$ represents the ground state of the interacting system. N_o is the number of bare particles with zero momentum at zero temperature. N_o will, in general, be considerably less than N, in view of the "depletion" of the zero momentum state as a consequence of particle interaction. (It will, nonetheless, correspond to macroscopic occupation of this state, being of order N, not of order 1.) This result is quite different from that of (2.14), which gives the *quasi-particle distribution* at $T = 0$. Throughout this chapter we shall be concerned exclusively with the latter quantity, aside from certain considerations of the microscopic theory in Chapters 9 and 10.

2.4 Density Fluctuation Excitation Spectrum

We consider next the spectrum of elementary excitations associated with the density fluctuations in the Bose liquid at $T = 0$. Just as for the non-interacting system, a density fluctuation of momentum $\mathbf{q}$ may be formed by exciting a quasi-particle of momentum $\mathbf{q}$, from the condensate. The corresponding excitation energy will be that of the quasi-particle, ε_q, and will be well-defined to the extent that the quasi-particle is well-defined.

There will be, in addition, contributions to the density fluctuation excitation spectrum from multi-particle configurations, involving two or more quasi-particles, of total momentum $\mathbf{q}$. In contrast to the single quasi-particle excitations, the multi-particle excitations will be spread over a continuous range of energies, as one allows the momentum of any single quasi-particle component to vary.

The spectrum of the density fluctuation excitations may be described in terms of the dynamic form factor, $S(\mathbf{q},\omega)$, introduced in Chapter 2, Vol. I. According to the above considerations, one may expect that $S(\mathbf{q},\omega)$ will have the form:

$$S(\mathbf{q},\omega) = NZ_q\delta\left(\omega - \varepsilon_q\right) + S^{(1)}(\mathbf{q},\omega) \tag{2.19}$$

The first term on the right-hand side of (2.19) represents a contribution of strength NZ_q to $S(\mathbf{q},\omega)$ from single quasi-particle excitations; the second is the contribution from configurations involving two or more

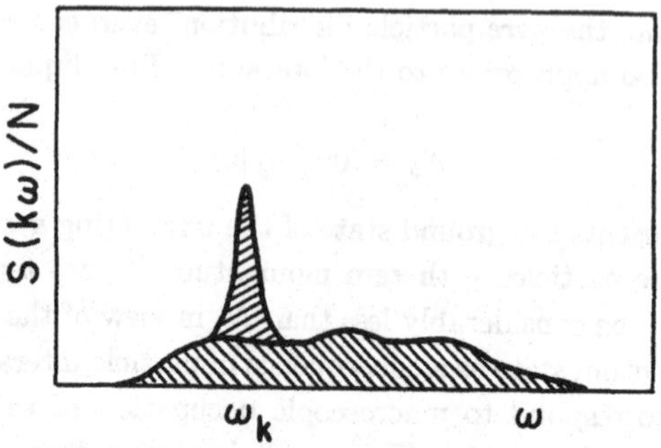

FIGURE 2.1. *Schematic representation of* $S(\mathbf{k}, \omega)$.

quasi-particles. A schematic representation of (2.19) is given in Fig. 2.1. In (2.19) we have assumed that the quasi-particle excitations are not damped. As we have seen, this represents something of an over-simplification, since the quasi-particles will, in general, possess a finite lifetime. An expression of the form (2.19) will represent a reasonable approximation when that lifetime is sufficiently long. We shall see that in the long wave-length limit this condition is well satisfied.

It was shown in Chapter 2, Vol. I, that the dynamic form factor for a neutral system must satisfy the following sum rules:

$$\int_0^\infty d\omega S(\mathbf{q}, \omega)\omega = \sum_n \left|(\rho_\mathbf{q}^+)_{no}\right|^2 \omega_{no} = \frac{Nq^2}{2m} \tag{2.20}$$

$$\lim_{\mathbf{q}\to 0} \int_0^\infty d\omega \frac{S(\mathbf{q}, \omega)}{\omega} = \lim_{\mathbf{q}\to 0} \sum_n \frac{\left|(\rho_\mathbf{q}^+)_{no}\right|^2}{\omega_{no}} = \frac{N}{2ms^2} \tag{2.21}$$

The first of these is the f-sum rule, the second, which governs only the long wave-length behavior of the dynamic form factor, is the compressibility sum rule.

Let us consider first the contribution to the sum rules, (2.20) and (2.21), from the multi-particle excitations. The arguments we use are identical in form to those developed in Chapters 2 and 4, Vol. I, for estimating the contribution to the dynamic form factor of a Fermi liquid from multi-pair excitations. Thus, since a given multi-particle state, $|n\rangle$, corresponds to a superposition of quasi-particles of net momentum $\mathbf{q}$, one expects no special $\mathbf{q}$ dependence of the corresponding excitation frequency, ω_{no}. Hence, in the limit $\mathbf{q} \to 0$, one finds

$$\lim_{\mathbf{q}\to 0} \omega_{no} \simeq \bar{\omega} \tag{2.22}$$

where $\bar{\omega}$ is some constant. Next we estimate the multi-particle matrix elements, $\left(\rho_{\mathbf{q}}^{+}\right)_{no}$. We note that for a translationally invariant system,

$$\lim_{\mathbf{q}\to 0} \left(\mathbf{J}_{\mathbf{q}}^{+}\right)_{no} = (\mathbf{J})_{no} = 0 \tag{2.23}$$

since the total current, $\mathbf{J}$, is a good quantum number. Hence,

$$\lim_{\mathbf{q}\to 0} \left(\mathbf{q} \cdot \mathbf{J}_{\mathbf{q}}^{+}\right)_{no} = q^{(1+a)} (a > 0) \tag{2.24}$$

Now, using longitudinal current conservation, (I.2.27), we can write:

$$\lim_{\mathbf{q}\to 0} \left(\rho_{\mathbf{q}}^{+}\right)_{no} = \lim_{\mathbf{q}\to 0} \frac{\left(\mathbf{q} \cdot \mathbf{J}_{q}^{+}\right)_{no}}{\omega_{no}} \sim \frac{q^{1+a}}{\bar{\omega}} \tag{2.25}$$

for an excited state, $|n\rangle$, which represents a multi-particle configuration. With the aid of (2.22) and (2.25), inspection of the sum rules, (2.20) and (2.21), reveals that the multi-particle configurations do not contribute in the long wave-length limit

It follows that in this limit the single quasi-particle excitations must exhaust *both* sum rules. On substituting (2.19) into (2.20) and (2.21) [and neglecting the contribution from $S^{(1)}(q, \omega)$], we find:

$$\lim_{q\to 0} \varepsilon_q = sq \tag{2.26}$$

$$\lim_{q\to 0} Z_q = \frac{q}{2ms} \tag{2.27}$$

$$\lim_{\mathbf{q}\to 0} S(\mathbf{q}, \omega) = \frac{Nq}{2ms} \delta(\omega - sq) \tag{2.28}$$

The results, (2.26) to (2.28), are very important. They tell us that the identical long wave-length quasi-particle and density-fluctuation excitations are phonons, with a velocity equal to the macroscopic sound velocity. This result is in agreement with the neutron scattering experiments mentioned in Chapter 1 which we shall discuss in further detail in Chapter 3.

It is, of course, straightforward to calculate the static form factor, $S_{\mathbf{q}}$, from (2.28). One finds

$$\lim_{\mathbf{q}\to 0} S_{\mathbf{q}} = \frac{q}{2ms} \tag{2.29}$$

We shall make use of this result in our discussion of the neutron scattering experiments on liquid He II.

The results we have obtained for the single quasi-particle and multi-particle contributions to the various matrix elements and sum rules of interest are summarized in Table 2.1. In estimating the multi-particle matrix elements, we have taken a [cf. (2.24)] to be unity, this being its most likely value in the absence of coherence effects associated with multi-particle configurations.

TABLE 2.1. *Matrix Elements, Excitation Frequencies, and Sum Rule Contributions from Single Quasi-Particle and Multi-Particle Excitations*

Quantity	Quasi-Particle	Multi-Particle ($a = 1$)
$\dfrac{(\mathbf{q}\cdot\mathbf{J}_q^+)_{no}}{q}$	$\left(\dfrac{Nqs}{2m}\right)^{1/2}$	q
$(\rho_\mathbf{q}^+)\, no$	$\left(\dfrac{Nq}{2ms}\right)^{1/2}$	q^2
ω_{no}	sq	$\bar{\omega}$
$\sum_n \omega_{no}\left\|(\rho_\mathbf{q}^+)_{no}\right\|^2$	$\dfrac{Nq^2}{2m}$	q^4
$\dfrac{\sum_n \left\|(\rho_\mathbf{q}^+)_{no}\right\|^2}{\omega_{no}}$	$\dfrac{N}{2ms^2}$	q^4
$\dfrac{\sum_n \left\|(\rho_\mathbf{q}^+)_{no}\right\|^2}{N}$	$\dfrac{q}{2ms}$	q^4

CHAPTER 3

ELEMENTARY EXCITATIONS IN HE II[1]

We have considered thus far the nature of the elementary excitation spectrum in an arbitrary Bose liquid. We now consider what this spectrum is like for the only Bose liquid found in nature, He II. As we have remarked in Chapter 1, indirect information on the excitation spectrum of He II can be obtained via a measurement of ρ_n and, as well, through measurements of the specific heat. If one assumes the Landau theory to be correct (and we shall see later that it is to within minor modifications), one can work backwards from the experimental data to a form of the quasi-particle spectrum consistent with that data. At the very lowest temperatures ($T \lesssim 0.5°K$) only long wave-length phonons are thermally excited. The experimental results for ρ_n and C_v are consistent with a phonon velocity equal to that of the measured macroscopic sound velocity, 237 m/sec. As the temperature increases, roton excitations become important; above 1°K they represent the dominant excitation mode. The roton contribution to ρ_n and C_v is consistent with the Landau excitation spectrum,

$$\varepsilon_{\text{roton}}(p) = \Delta + \frac{(p - p_o)^2}{2\mu} \tag{1.8}$$

In Table 3.1 we list various experimental determinations of the parameters of the roton spectrum; it may be seen that more accurate thermodynamic and second sound measurements have led to a better determination of these parameters.

[1]Much of the treatment in this chapter parallels closely that given in Miller, Pines and Nozières (1962).

TABLE 3.1. *Parameters of the Roton Spectrum [from J. de Boer, Proceedings of the Intl. School of Physics at Varenna, Course 21 (Academic Press, New York, 1963), p. 1.]*

Source	Investigators		Δ/k (°K)	$p_o/\hbar\,(A^{-1})$	μ/m_{He}
Thermodynamic data, second sound	Landau[a]		9.6	1.95	0.77
	Khalatnikov[b]		8.9 ± 0.2	1.99 ± 0.05	0.26 ± 0.09
	Wiebes et al.[c]		8.8	1.96	0.23
Neutron scattering	Yarnell et al.[d]	1.1°K	8.65 ± 0.04	1.92 ± 0.01	0.16 ± 0.01
		1.6°K	8.43	1.92	0.16
		1.8°K	8.15	1.92	0.16
	Henshaw and Woods[e]	1.8°K	8.65 ± 0.11	1.91 ± 0.01	0.16
		1.8°K	8.1	1.91	0.16
	Palevsky et al.[f]		8.1		

[a] L. D. Landau, *J. Phys. USSR* **11**, 91 (1947).
[b] I. M. Khalatnikov, *Fortschr. d. Phys.* **5**, 211 (1957).
[c] J. Wiebes, C. G. Niels-Hakkenberg, and H. C. Kramers, *Physica* **23**, 625 (1957).
[d] J. L. Yarnell, G. P. Arnold, P. J. Bendt, and E. G. Kerr, *Phys. Rev.* **113**, 1379 (1959).
[e] D. G. Henshaw and A. D. B. Woods, *Phys. Rev.* **121**, 1266 (1961).
[f] H. Palevsky, K. Otnes, and K. E. Larsson, *Phys. Rev.* **112**,11 (1958).

Measurements of ρ_n and C_v provide us with information on only a small part of the quasi-particle energy vs. momentum curve. Even at 2°K only long wave-length phonons ($p \lesssim 0.1$ Å^{-1}) and rotons with momenta in the immediate vicinity of p_o are thermally excited. By contrast, neutron scattering experiments furnish us with data for the quasi-particle spectrum over essentially the entire range of momenta for which the quasi-particles are well-defined. We therefore turn to a consideration of the available neutron scattering data.

3.1 Neutron Scattering

We have seen in Sec. 2.1, Vol. I, that at $T = 0$ a measurement of the angular distribution of a beam of inelastically scattered neutrons furnishes a direct measure of the dynamic form factor, $S(\mathbf{q}, \omega)$, for a many-particle system. Let us, for simplicity, assume that the He II system is in its

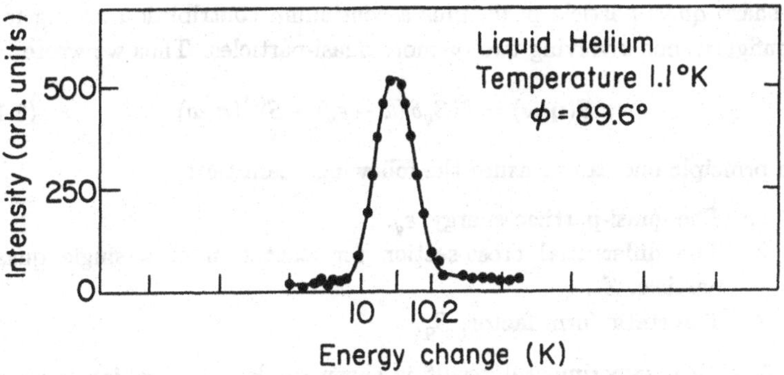

FIGURE 3.1. *Inelastic scattering cross-section for neutrons (at 89.6°) [from Henshaw and Woods (1961)].*

ground state. (The generalization of our results to finite temperatures is straightforward, and will be considered in Chapter 7.) In Chapter 2 we saw that $S(\mathbf{q}, \omega)$ for a Bose liquid will, in general, consist in two parts:

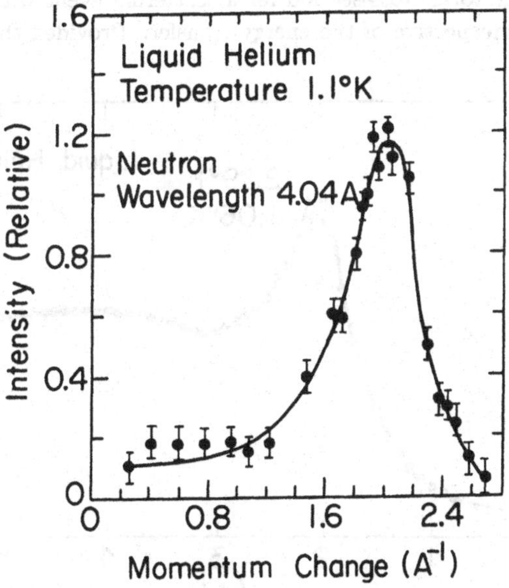

FIGURE 3.2. *Differential cross-section for single quasi-particle production [Henshaw and Woods (1961)].*

a sharp quasi-particle peak, plus a continuum contribution arising from configurations involving two or more quasi-particles. Thus we wrote:

$$S(\mathbf{q},\omega) = NZ_q\delta(\omega - \varepsilon_q) + S^{(1)}(\mathbf{q},\omega) \tag{2.19}$$

In principle one can measure the following quantities:

1. The quasi-particle energy, ε_q.
2. The differential cross-section for excitation of a single quasi-particle, Z_q.
3. The static form factor, $S_\mathbf{q}$.

A typical experimental result is shown in Fig. 3.1, which gives the number of neutrons which have suffered an energy loss ω whilst being scattered through an angle θ. The peak in the curve corresponds to those neutrons which have excited a single quasi-particle from the condensate; its position yields ε_q, while its strength (i.e., the area enclosed) gives Z_q. In Fig. 1.6 we have reproduced the experimental results at $T = 1.1°\text{K}$ of Henshaw and Woods (1961) for ε_q, while in Fig. 3.2 we give their corresponding plot for Z_q.

As we have seen in Sec. 2.1, Vol. I, the static form factor, $S_\mathbf{q}$, corresponds to the total cross-section for a scattering event with momentum transfer $\mathbf{q}$, irrespective of the energy transfer. Provided the momentum

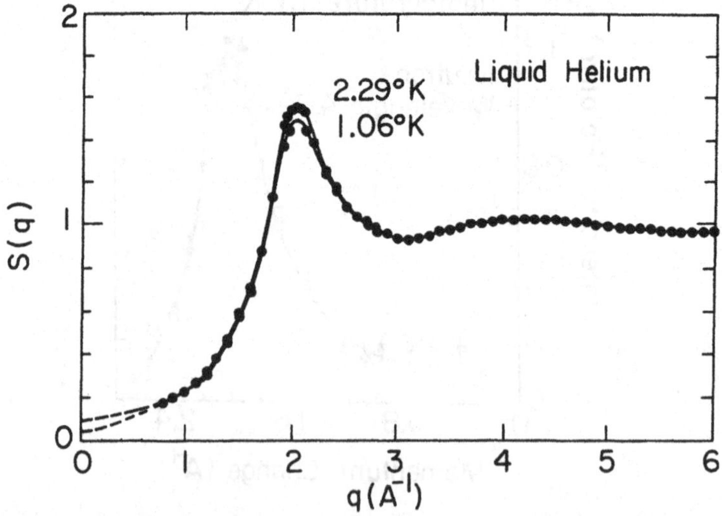

FIGURE 3.3. *Experimental results for $S_\mathbf{q}$ [Henshaw (1960)].*

of the incident neutron is sufficiently large, it may be obtained from a measurement of the scattered intensity at a fixed angle, θ. In Fig. 3.3 we reproduce the experimental results for $S_{\mathbf{q}}$ obtained by Henshaw (1960).

3.2 Theoretical Analysis

We first consider in more detail the measured excitation spectrum. As we might expect from our $T = 0$ analysis, in the limit of very small $\mathbf{q}$ the spectrum is phonon-like, with

$$\varepsilon_q = sq \tag{3.1}$$

where s agrees well with the experimentally measured macroscopic sound velocity.[2] What is perhaps surprising is that the linear portion of the ε vs. q curve extends up to $q \sim 0.6$ Å^{-1}, a wave-vector which is enormous compared to those macroscopic values of q for which the analysis of Chapter 2 is obviously satisfactory.

We next note that the ε vs. q curve varies smoothly from $q = 0$ to $q = 2.68$ Å^{-1}, where it cuts off abruptly. For small q, the curvature is downward, so that it is not possible for one phonon to decay into two (with conservation of energy and momentum). Indeed throughout the entire observed excitation curve, it is not possible for a single quasi-particle excitation to decay into two quasi-particles of lower momentum. It is for precisely this reason that the quasi-particle excitations are well-defined over such a substantial momentum region. The cut-off in the quasi-particle spectrum occurs at the first momentum, $\mathbf{p}_c$, at which such a decay becomes energetically possible. The actual mechanism is that of decay into two rotons, since the energy of a quasi-particle of momentum 2.68 Å^{-1} is very nearly $2\Delta \simeq 17.3$°K.

[2]The alert reader may object to our using a $T = 0$ analysis at a finite temperature, where collisions may act to change the density-fluctuation and quasi-particle excitation spectrum. The modifications which arise in the long wave-length limit will be discussed in Chapter 7; here we remark that the $T = 0$ analysis is essentially a "collisionless" analysis, which may be expected to hold true at a temperature T provided the momentum q is such that $\varepsilon_q \tau \gg 1$ where τ is a characteristic collision time for the excitation in question. At 1°K, $\tau \sim 10^{-9}$ sec, so that the $T = 0$ analysis will apply to that part of the spectrum for which $q \gg 10^5$ Å^{-1}. Since the lowest momentum excitation measured by Henshaw and Woods is $\sim 0.6 \times 10^7$, one is obviously in the collisionless regime.

Pitaevskii (1959, 1961a) has carried out an elegant field-theoretic analysis of such an endpoint to the quasi-particle spectrum. He has shown that near the endpoint, the energy of a quasi-particle should be

$$\varepsilon_p \simeq 2\Delta - \alpha \exp\left[-\frac{a}{|\mathbf{p}-\mathbf{p}_c|}\right] \quad (p \lesssim \mathbf{p}_c) \tag{3.2}$$

where p_c is the momentum for which the decay first becomes possible, and α and a are numerical constants. The experimental measurements by Henshaw and Woods are consistent with an endpoint spectrum of this form, though precise measurements are required to fit the constants α and a.

We next consider the results obtained by Henshaw and Woods for Z_q, the differential cross-section for the production of a single quasi-particle. The results shown in Fig. 3.2 represent relative measurements of this quantity; in order to convert them to absolute measurements it is necessary that one find an independent scheme for defining the scale of Z_q. This we may do with the aid of the long wave-length sum rule considerations of Chapter 2. There we saw that for very small $\mathbf{q}$,

$$\lim_{\mathbf{q}\to 0} Z_q = S_{\mathbf{q}} = \frac{q}{2ms}$$

If now we use the observed values for the sound velocity, the normalization is easily accomplished by putting a straight line of the desired slope through the points at 0.2 Å^{-1} and 0.4 Å^{-1} and the origin. The linear behavior of the Z_q curve up to such large values of q is consistent with the fact that the ε_q curve remains linear up to $q \sim 0.6$ Å^{-1}. The resulting curve is shown in Fig. 3.4.

We turn next to a consideration of Henshaw's measurement of $S_{\mathbf{q}}$. The experimental results extend down to wave-vectors of the order of 0.8 Å^{-1}. Again, one may attempt to extrapolate these into the lower momentum region with the aid of (2.29). If one does this, one finds that the resulting straight line does not join smoothly on to the $S_{\mathbf{q}}$ curve for $q \gtrsim 0.8$ Å^{-1}. If we assume the normalization of the $S_{\mathbf{q}}$ curve to be essentially correct, then we must conclude that the correct curve for $S_{\mathbf{q}}$ has a form not too different from that shown in Fig. 3.5. There we see a linear behavior up to 0.4 Å^{-1} (consistent with the normalization of Z_q) followed by a mild shoulder between 0.4 Å^{-1} and 0.8 Å^{-1}. It will be interesting to see whether future experiments indicate such behavior in this momentum region.

With the aid of the normalized curves for Z_q and $S_{\mathbf{q}}$ we can answer the following question: at what wave-vectors does $S^{(1)}(\mathbf{q}, \omega)$, the contribution

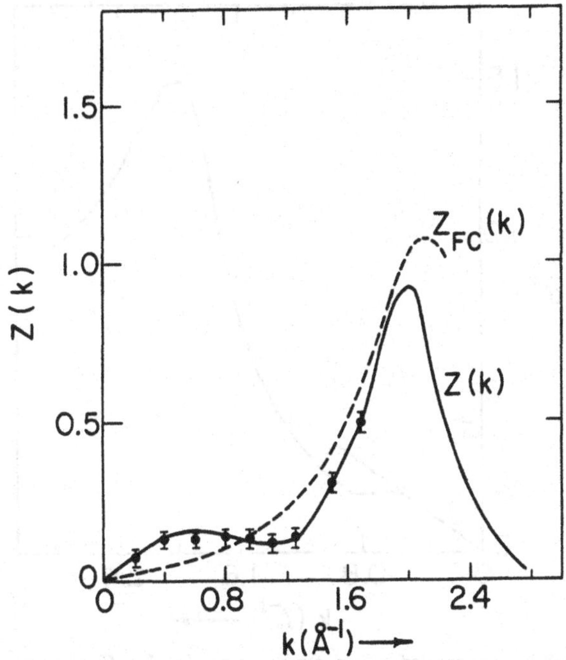

FIGURE 3.4. *Experimental [$Z(k)$)] and theoretical [$Z_{FC}(k)$ Cohen and Feynman (1957)] for $Z(k)$.*

to $S(\mathbf{q},\omega)$ from excitations involving two or more quasi-particles, begin to play an important role? We saw in Chapter 2 that for a translationally invariant system $S^{(1)}(\mathbf{q},\omega)$ must vanish in the long wave-length limit. On substituting (2.19) into the equation which defines $S_\mathbf{q}$, (I.2.18), we see that $S^{(1)}(\mathbf{q},\omega)$ must satisfy the moment equation:

$$\int_0^\infty d\omega\, S^{(1)}(\mathbf{q},\omega) + NZ_q = NS_\mathbf{q} \tag{3.3}$$

It is convenient to divide this equation by $NS_\mathbf{q}$ and to introduce the ratio

$$f_q = \frac{Z_q}{S_\mathbf{q}} \tag{3.4}$$

We then find

$$\int_0^\infty d\omega\, \frac{S^{(1)}(\mathbf{q},\omega)}{NS_\mathbf{q}} = 1 - f_q \tag{3.5}$$

We see that f_q provides a direct measure of the contribution made by the

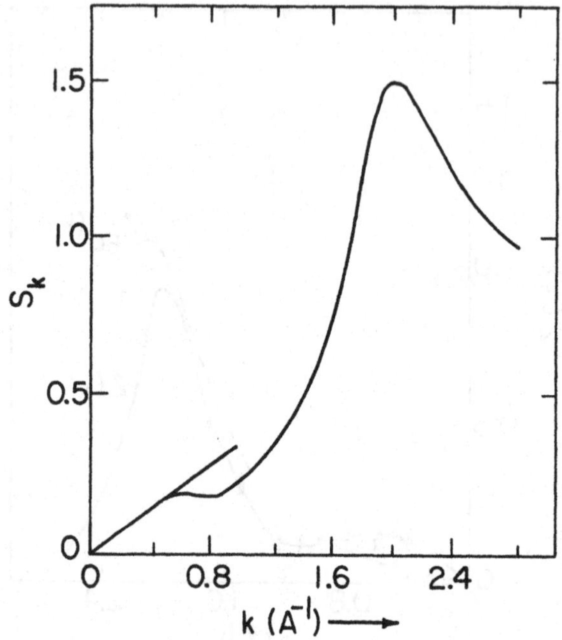

FIGURE 3.5. *Suggested static form factor for He II [Miller, Pines and Nozières (1962)].*

multi-particle excitations to the dynamic form factor; it represents the relative efficiency of single quasi-particle excitation from the condensed state. A plot of f_q, based on the experimental measurements (and our extrapolations thereof), is given in Fig. 3.6. We note that $f_q \simeq 1$ for values of q extending from the origin to 0.4 Å^{-1}; this approximate equality is a direct consequence of our extrapolations of the experimental data. Beyond 0.4 Å^{-1}, f_q begins to fall off, as the multi-particle excitations begin to contribute to $S(\mathbf{q}, \omega)$. At first sight we might expect that this fall-off would continue in some smooth fashion out to the largest momentum values, with multi-particle excitations playing an increasingly important role. In fact, there must be a second physical effect, which is responsible for the maximum of f_q in the roton region. We postpone discussion of the effect until after we have discussed the nature of rotons in more detail.

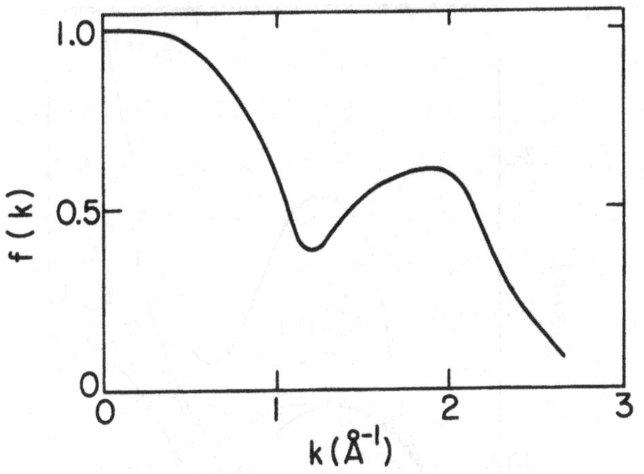

FIGURE 3.6. *Relative efficiency of single quasi-particle production [Miller, Pines and Nozières (1962)].*

3.3 Feynman Excitation Spectrum

We consider now the extent to which theory has been able to account for the elementary excitation spectrum in He II. The calculations we shall describe are essentially phenomenological and variational in their nature; there do not at present exist any satisfactory microscopic calculations.

The first detailed calculation of the excitation spectrum was carried out by Feynman (1955). Feynman argued that one could not easily calculate the ground state wave-function, but that one should be able to write down an excited state wave-function, and then obtain the corresponding excitation energy in terms of measured properties of the liquid. Feynman was led, by means of a series of elegant physical arguments (for which we refer the reader to his classic paper), to the following excited-state wave-function [which had earlier been considered by Bijl (1940)]:

$$\psi_F = \rho_{\mathbf{q}} \psi_o \tag{3.6}$$

where ψ_o is the exact ground state wave-function for the liquid. He showed that the corresponding excitation spectrum would be given by

$$\epsilon_q^F = \frac{q^2}{2mS_q} \tag{1.9}$$

where S_q is the measured static form factor.

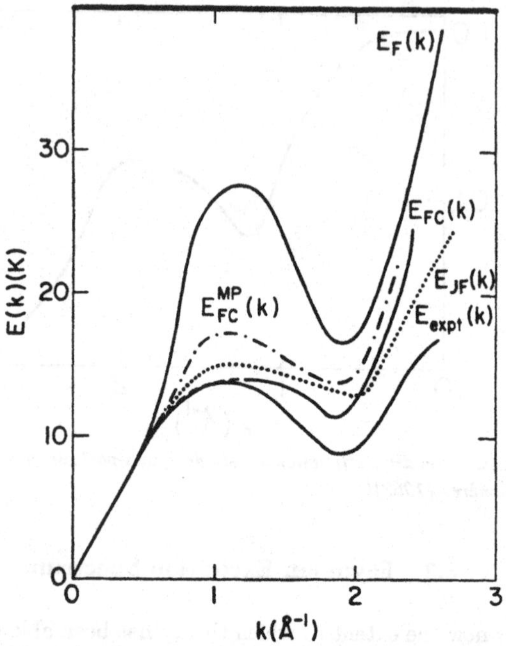

FIGURE 3.7. *Comparison of Jackson and Feenberg (1962) variational calculation of the excitation spectrum of liquid He II with prior calculations and with experiment.*

The Feynman result, (1.9), may be derived quite simply with the aid of the exact moment equations for $S(\mathbf{q}, \omega)$, (2.20) and (3.3). Use of the wave-function (3.6) as an eigenstate of the liquid implies that the density fluctuation excitation spectrum consists in a single excited state of momentum **q**; this state is, of course, the single quasi-particle contribution to $S(\mathbf{q}, \omega)$. On neglecting the multi-particle excitation contribution, $S^{(1)}(\mathbf{q}, \omega)$, to (2.20) and (3.3) *for all values of* **q**, we find, from (3.3),

$$Z_q = S_{\mathbf{q}} \tag{3.7}$$

while the *f*-sum rule (2.20) takes the form

$$NZ_q\epsilon_q = \frac{Nq^2}{2m} \tag{3.8}$$

whence (1.9) follows at once.

When the experimental values of $S_{\mathbf{q}}$ are substituted into (1.9), the resulting excitation energy is in good qualitative agreement with experiment, as shown in Fig. 3.7. We have seen that the Feynman wave-function must be *exact* in the long wave-length limit. In fact, it gives a quantitative account of the spectrum from $q \sim 0$ to $q \sim 0.6$ Å^{-1}, that is throughout nearly the entire linear portion of the spectrum. It has, moreover, qualitatively the correct shape, showing both a maximum in the region of $q \sim 1$ Å^{-1} and a minimum at the roton momentum. The idea that a single quasi-particle excitation provides the dominant contribution to the density-fluctuation excitation spectrum is thus seen to be a good first approximation for the description of the excitations throughout the entire momentum range of interest.

3.4 Feynman-Cohen Theory

Feynman and Cohen (1956) attacked the problem of obtaining a better description of the quasi-particle excitations in the following way. They argued that the wave-function, (3.6), could not in general provide a correct description because it does not offer a proper account of the motion of an excitation through the liquid. If one localizes the excitation, (3.6), by construction of a wave-packet, one finds that the corresponding current is localized and equal to $\mathbf{q}/m$. Such a description cannot be appropriate for a roton at the minimum of the excitation spectrum, which has zero group velocity and corresponds to an excitation which is stationary in space: the corresponding current must necessarily vanish. Feynman and Cohen point out that what is missing from the description is the return flow of the rest of the liquid as the excitation moves through it; this *backflow* must be such as to bring about overall current conservation for the motion of a localized excitation.

We have already encountered the idea of backflow in Chapters 2 and 3, Vol. I. There we saw that accompanying the motion of a slowly-moving impurity atom in a fermion system there will be a backflow which, at large distances, takes a dipolar form. The *size* of the backflow depends on the dynamical properties of the system in question; for a charged fermion system, it is just such as to cancel the current carried by the impurity particle. Our study of backflow was, in fact, inspired by the work of Feynman and Cohen, who considered the motion of an impurity atom of equal mass, but not subject to Bose statistics, through liquid He II. If the impurity possesses a momentum $\mathbf{q}$ and is localized at some position, $\mathbf{R}$, a wave-function which allows for backflow takes the form:

$$\psi = \exp[-i\mathbf{q}\cdot\mathbf{R}]\exp\left[i\sum_i g(\mathbf{r}_i - \mathbf{R})\right] \tag{3.9}$$

At large distances, $g(\mathbf{r})$ possesses the dipolar form,

$$g(\mathbf{r}) = A\frac{\mathbf{q}\cdot\mathbf{r}}{r^3} \tag{3.10}$$

A variational calculation based on (3.10) shows that the energy of the impurity is minimized for a value of A very close to the "perfect screening" value,

$$A = -\frac{1}{4\pi N} \tag{3.11}$$

The result, (3.11), suggests that an impurity in liquid He, interacting via the same law of interaction as the He atoms, behaves very much like a charged impurity moving through the electron gas; it acts to induce a backflow in the liquid which effectively cancels the longitudinal part of the impurity current at large distances.

This result led Feynman and Cohen to propose a symmetrized version of (3.9) as a variational wave-function for the determination of the energy of an excitation in liquid He. Their wave-function was

$$\psi = \sum_i e^{i\mathbf{q}\cdot\mathbf{r}_i}\exp\left[\sum_{i\neq j} g(\mathbf{r}_i - \mathbf{r}_j)\right]\psi_o \tag{3.12}$$

where $g(\mathbf{r})$ was assumed to possess the dipolar form (3.10). In order to carry out their variational calculation of the excitation energy appropriate to (3.12), Feynman and Cohen found it necessary to expand the backflow term. Their calculation was thus based on the following wave-function:

$$\psi_{FC} = \sum_i e^{i\mathbf{q}\cdot\mathbf{r}_i}\left\{1 + iA\sum_{j\neq i}\frac{\mathbf{q}(\mathbf{r}_i - \mathbf{r}_j)}{|\mathbf{r}_i - \mathbf{r}_j|^3}\right\}\psi_o \tag{3.13}$$

The resulting value of the excitation energy (shown in Fig. 3.7) is in considerably better agreement with experiment than is the Feynman value derived from (1.9). Subsequently Cohen and Feynman (1957) used to wave-function (3.13) to calculate Z_q with the result shown in Fig. 3.4. We see that the calculation of both ε_q and Z_q agrees rather well with experiment for values of q extending from 0 to slightly beyond the roton minimum; for larger values of q the agreement is less satisfactory.

Let us consider the Feynman-Cohen theory from another point of view. It is clear from the above discussion that once the multi-particle excitation contributions to $S(\mathbf{q}, \omega)$ begin to play an appreciable role, the simple Feynman result, (1.9), can no longer be expected to be applicable. Backflow must correspond in some sense to taking into account the effect of such multi-particle configurations on the quasi-particle excitation spectrum. Indeed, $S^{(1)}(\mathbf{q}, \omega)$ originates in the coupling of $\rho_{\mathbf{q}}^{+}\psi_o$ to higher configurations involving several elementary excitations. If we view $\rho_{\mathbf{q}}^{+}\psi_o$ as a "bare" quasi-particle, the coupling expresses the physical fact that such a bare quasi-particle is surrounded by a cloud of virtual elementary excitations. As it moves the quasi-particle drags the cloud along with it; the cloud represents the backflow which guarantees current conservation. In the limit of $q \to 0$, $S^{(1)} \to 0$; the self-energy cloud plays no role, and one recovers the Feynman result for ε_q.

From this point of view, the variational approach of Feynman and Cohen corresponds to taking into account the interaction between Feynman quasi-particles described by the wave-function, (3.6). The correspondence is made evident if we Fourier-analyze the wave-function (3.13) which was actually employed by Feynman and Cohen. We find

$$\psi_{FC} = \left\{ \rho_{\mathbf{q}}^{+} + \sum_{\mathbf{k} \neq \mathbf{q}} 4\pi A \frac{\mathbf{q} \cdot \mathbf{k}}{k^2} \rho_{\mathbf{k}}^{+} \rho_{\mathbf{q}-\mathbf{k}}^{+} \right\} \psi_o \tag{3.14}$$

The Feynman-Cohen quasi-particle thus contains a finite admixture of configurations which contain *two* Feynman quasi-particles. Conversely, the operator $\rho_{\mathbf{q}}^{+}$ will couple the ground state ψ_o to excited states containing two or more Feynman-Cohen quasi-particles. The last term of (3.14) thus gives rise to the lowest-order multi-particle contribution to $S^{(1)}(\mathbf{q}, \omega)$.

Kuper (1955), prior to the work of Feynman and Cohen, had carried out a closely related calculation, in which he treated the interaction between the virtual Feynman quasi-particle excitations by means of perturbation theory. It is instructive to derive Kuper's wave-function, and to compare it with (3.14). To do so, we define a Hamiltonian, H^F, which yields directly the Feynman wave-function when acting on the ground state.

$$H^F \rho_{\mathbf{q}}^{+}\psi_o = (\varepsilon_F + E_o)\, \rho_{\mathbf{q}}^{+}\psi_o \tag{3.15}$$

where E_o is the ground state energy. Let us regard the difference between the true Hamiltonian, H, and H^F as a small perturbation. Using

Rayleigh-Schrödinger perturbation theory we then find that the first-order system wave function is given by

$$\left\{\left(\frac{\rho_{\mathbf{q}}^{+}}{NS_{\mathbf{q}}}\right)+\frac{1}{2}\sum_{\substack{\mathbf{k}\neq 0\\ \mathbf{k}\neq\mathbf{q}}} A_{\mathbf{k},\mathbf{q}}\rho_{\mathbf{q}-\mathbf{k}}^{+}\frac{\rho_{\mathbf{k}}^{+}}{N\left(S_{\mathbf{k}}S_{\mathbf{q}}\right)^{1/2}}\right\}\psi_o \tag{3.16}$$

where

$$A_{\mathbf{k},\mathbf{q}}=\frac{\langle\psi_o|\rho_{\mathbf{q}-\mathbf{k}}\rho_{\mathbf{k}}\left(H-H^F\right)\rho_{\mathbf{q}}^{+}|\psi_o\rangle}{\varepsilon_F(\mathbf{q})-\varepsilon_F(\mathbf{q}-\mathbf{k})-\varepsilon_F(\mathbf{k})} \tag{3.17}$$

and small overlap integrals have been neglected. An alternative expression for the numerator of $A_{\mathbf{k},\mathbf{q}}$ is the following:

$$\langle\psi_o|\rho_{\mathbf{q}-\mathbf{k}}\rho_{\mathbf{k}}\left\{\left[H,\rho_{\mathbf{q}}^{+}\right]-\varepsilon_F(q)\rho_{\mathbf{q}}^{+}\right\}|\psi_o\rangle \tag{3.18}$$

For small values of $\mathbf{q}$, we have

$$\lim_{\mathbf{q}\to 0}\left[H,\rho_{\mathbf{q}}^{+}\right]=\varepsilon_F(q)\rho_{\mathbf{q}}^{+} \tag{3.19}$$

The strength of the perturbation, $(H-H_F)$, then vanishes. As we go to larger values of $\mathbf{q}$, the relation (3.19) is no longer valid, and the perturbation is non-vanishing.

Kuper used (3.16) to calculate the energy of a roton (of momentum $\mathbf{p}_o$). He found $E_r \simeq 11.5°\mathrm{K}$, a value which is in good agreement with that obtained by Feynman and Cohen. Subsequently Jackson and Feenberg (1962) carried out a calculation similar to that of Kuper's but in which Brillouin-Wigner perturbation theory was used in place of Rayleigh-Schrödinger theory (the energy denominators in (3.17) are replaced by the observed excitation energies). The Jackson-Feenberg results, which extend over the entire momentum region, are likewise shown in Fig. 3.7. The close agreement with the results of Feynman and Cohen is gratifying, but not surprising, in view of the close connection between (3.23) and (3.21). Indeed, as Miller and the authors have argued, it should be possible to show that (3.16) reduces to (3.14) in the limit of small $\mathbf{q}$, since in this limit the admixture of two quasi-particle excitations should be given accurately by perturbation theory.

3.5 What is a Roton?

The good agreement between the various theoretical calculations of the roton energy, and that observed experimentally, leads one to conclude that either of the wave-functions (3.14) or (3.16) provides a good description of that excitation. We may then inquire as to what sort of a physical picture of the excitation the wave functions offer. One picture, which has been espoused by Feynman, is that the roton is a vortex ring of such small radius that only a single atom can pass through its center. The backflow corresponds to a slow drift of atoms outside the ring which are returning for another passage through it.

A somewhat different picture of roton motion has been proposed by Miller, Pines and Nozières (1962). If one carries out in the Bogoliubov approximation a microscopic calculation of the coupling of an impurity atom to the phonons in a Bose gas, one finds a wave-function of exactly the Feynman-Cohen type. The backflow acts to cancel the impurity current; moreover it increases the effective mass of the impurity to

$$m_I^* = \frac{m_I}{(1-\alpha)}$$

$$\alpha = \frac{N'}{3N}\frac{m}{m_I} \tag{3.20}$$

where m_I is the mass of the impurity atom and N' is the number of phonon modes one has taken into account. It is tempting to speculate that this result, being independent of the specific parameters of the problem, is true for a Bose liquid as well. If further, one argues that the number of phonon modes which contribute to the backflow is maximum, and equal to N, then one finds that

$$m^* = \frac{3}{2}m \tag{3.21}$$

for an impurity of mass equal to that of ^{4}He. The result (3.21) is likewise found for an entirely different problem; it is the classical hydrodynamic effective mass of a sphere moving through a liquid.

If, now, we regard the Feynman excitations, (3.6), of wave-vector greater than 1 Å^{-1}, say, as being coupled to the phonons in much the same way as is an impurity atom, we would expect that the result of including backflow would be to alter the energy to

$$E^*(q) = \frac{q^2}{2m^* S_q} = \frac{q^2}{3mS_q} \tag{3.22}$$

The result (3.22) provides, in fact, a good approximation to the Feynman-Cohen calculation throughout the momentum region 1 Å^{-1} to 2 Å^{-1}. (We chose $q > 1$ Å^{-1}, because for smaller wave-vectors we may expect coherence factors to reduce markedly the coupling between the various Feynman excitations.) We note that in the roton region, the Feynman excitation is itself very nearly a free particle; it differs slightly in that it possesses a 30% higher effective mass, the result one might say of interaction with its nearest neighbors. We may than argue that a roton is a slightly modified free particle excitation, which moves surrounded by a cloud of phonons, the latter describing the backflow of the other atoms at long distances from the atom in question.

This view, which has been put forth in slightly different language by Chester (1963), is not so different from that of Feynman as it might at first seem. For, as Chester has pointed out, the velocity potential for a classical smoke ring is of just the backflow type. It is then a matter of taste whether one wishes to picture the roton as the smallest possible vortex ring or as a modified free particle excitation. In both cases the flow pattern far from the atom is of the dipolar form we have considered; in both cases one does not have an altogether accurate description of what is happening near the center of the excitation. It is this lack of knowledge of the behavior of the system at distances of the order of the inter-particle spacing which has rendered difficult the construction of an accurate description of excitations in the momentum region, 2 Å^{-1} to 2.7 Å^{-1}. Here the long wave-length approximation (or what is equivalent to the two quasi-particle approximation) implicit in (3.14), is no longer valid; moreover, the idea of a purely dipolar backflow is no longer applicable.

In conclusion we return to the question of the physical effect which is responsible for the peak in f_q in the region of the roton momentum. A possible explanation of this maximum has been suggested by J. Bardeen. It is based on the fact that in the vicinity of the roton momentum, the excitations have a very small or vanishing group velocity. As a result, there is "time," so to speak, to form the excitation with its appropriate backflow of other virtual excitations. As one goes away from the roton minimum in either direction, the group velocity increases. The coupling to the appropriate "dressed" excitation (one with the right backflow of other virtual excitations) is correspondingly less efficient until one reaches a momentum region in which such coherence effects play no role. We note that in the region between 1 Å^{-1} and 2 Å^{-1} the Feynman-Cohen wave-function offers a correct description of this phenomenon too, since their calculation of Z_q is in accord with experiment here.

3.6 Quasi-Particle Excitations at Finite Temperatures

In our discussion in Chapter 2, we remarked that at finite temperatures one might expect the interaction between thermally-excited quasi-particles to influence the quasi-particle excitation spectrum in two ways:

1. The quasi-particle energy will vary with temperature. Such variation may be expected to be negligible as long as the relative number of thermally-excited quasi-particles is negligible, that is, when

$$\frac{N'}{N} = \sum_{p}^{p\neq 0} \frac{n_p(T)}{N} \ll 1 \tag{3.23}$$

2. There will be an "uncertainty-principle" broadening of the quasi-particle energy, as a result of collisions between the thermally-excited quasi-particles. This effect likewise becomes increasingly important as T approaches T_c.

Neutron scattering experiments which measure both the shift and the width of quasi-particle energies with temperature have been carried out at several temperatures for excitations in the vicinity of the roton minimum. We report briefly on the results here.

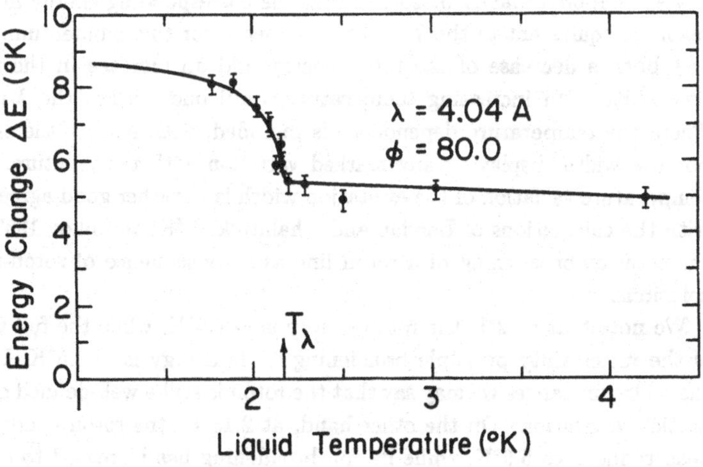

FIGURE 3.8. *Temperature variation of excitation energy (near roton minimum) as measured by Henshaw and Woods (1961).*

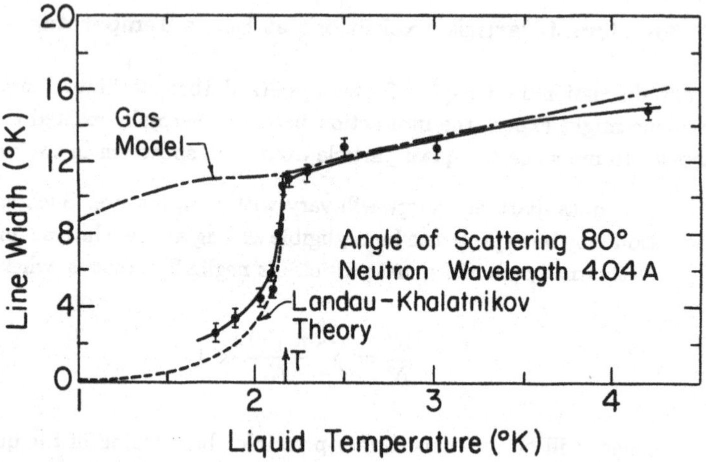

FIGURE 3.9. *Temperature variation of line width (near roton minimum) as measured by Henshaw and Woods (1961).*

In Figs. 3.8 and 3.9, we reproduce the experimental results of Henshaw and Woods (1961) for the energy and line width of rotons which are effectively at the minimum of the excitation curve ($p \sim 1.91$ Å^{-1}). We note that at 1.1°K the relative number of thermally excited quasi-particles, N'/N, is approximately 0.25%, so that the corresponding energy may be taken as equivalent to the $T = 0$ roton energy for this momentum. One sees both a decrease of the roton energy and an increase in the roton line width with increasing temperature, until one reaches the λ-point. There the temperature dependence is modified, with neither the energy nor the width displaying any marked variation with temperature. The temperature variation of the excitation width is in rather good agreement with the calculations of Landau and Khalatnikov (Khalatnikov 1957) for the collision-broadening of a roton line as a consequence of roton-roton collisions.

We note that at 2°K the roton energy is $\sim$ 7.5°K, while the full width of the uncertainty principle broadening of its energy is $\sim$ 4.5°K. Under these circumstances we may say that the roton is still a well-defined quasi-particle excitation. On the other hand, at 2.13°K, the roton energy has been reduced to 5.9°K, while its full broadening has increased to 8.0°K. Clearly in the immediate vicinity of the λ-point, rotons cease to be well-defined elementary excitations in the conventional sense ($\varepsilon_p \tau \gtrsim 1$, say).

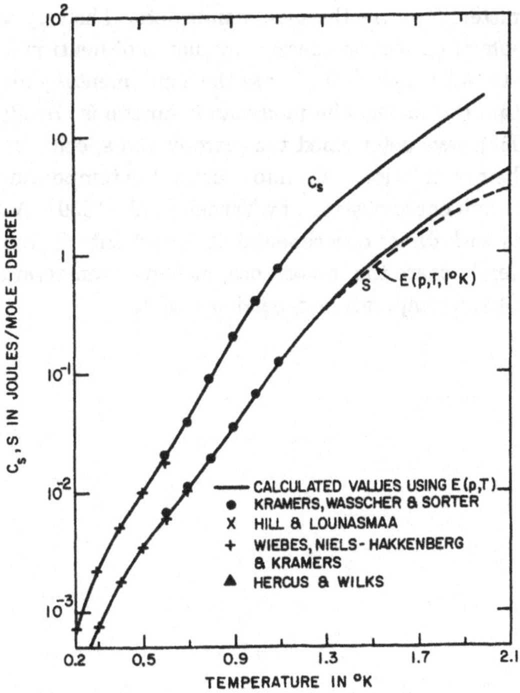

FIGURE 3.10. *Comparison of calculated and experimental values of entropy and specific heat of liquid helium II. The dashed line was calculated disregarding the temperature dependence of the excitation curve [from Bendt, Cowan, and Yarnell (1959)].*

The shift in excitation energies is not confined to the roton minimum; earlier measurements by Yarnell et al. (1959) for excitations with momenta between 1.6 Å^{-1} and 2.1 Å^{-1} show a similar decrease of excitation energy with temperature over the entire range of momentum and temperature (1.1°K to 1.8°K) studied.

There exists as well, from ultrasonic attenutation measurements, experimental information on the temperature variation of quasi-particle energies in a region of much lower momentum ($p \sim 10^2$ cm^{-1}) and temperature ($T < 0.7$°K). In this "phonon" regime, the temperature variation of ε_p is rather different from the roton region considered here; it will be discussed in Chapter 7.

We have seen that experimental study of the elementary excitation spectrum in He II effectively began with Landau's determination of the roton parameters from the thermodynamic data. The very accurate direct measurements of excitation energies by means of neutron scattering now permit one to work backwards; to use the experimentally measured values of ε_p to obtain the various thermodynamic functions. Bendt, Cowan, and Yarnell (1959) have determined the entropy and specific heat of He II in this way; their calculations take into account the temperature dependence of the excitation curve observed by Yarnell et al. (1959). A comparison of their results with direct experimental measurements is given in Fig. 3.10. One sees clearly there the importance, at higher temperatures, of taking into account the temperature dependence of ε_p.

CHAPTER 4

SUPERFLUID BEHAVIOR: RESPONSE TO A TRANSVERSE PROBE. QUALITATIVE BEHAVIOR OF A SUPERFLUID

The difference between normal systems and superfluid systems is perhaps most strikingly manifested in their current flow. Superfluid current flow displays very special features, both in its structure and in its response to external probes. We now focus our attention on these characteristic "superfluid" properties at zero-temperature. The generalization of the theory to finite temperatures is postponed until Chapter 6.

Two fundamental features of superfluid flow are the following:

1. Superfluid flow is *irrotational*. This property may be viewed in a number of ways. In the language of a two-fluid model, it is expressed by the equation:

$$\nabla \times \mathbf{v}_s = 0$$

where $\mathbf{v}_s$ is the superfluid velocity, which we shall define more precisely in Chapter 5. More conveniently, we may say that it is impossible to set up rotational currents on a macroscopic scale in a superfluid. It follows that when a *transverse probe* is applied to the fluid, the latter does not respond. For example, let us imagine a simple experiment, in which a bucket of helium is rotated at a constant angular velocity ω. In a frame of reference which rotates along with the walls, the fluid feels a Coriolis force, which acts as a transverse probe applied to the system. A normal liquid would respond to such a probe; at equilibrium, the liquid would be dragged by the walls and rotate at the same angular velocity ω. In a homogeneous superfluid at $T = 0$, the liquid does not respond to the

Coriolis force; it stays at rest while the walls are rotating. In an actual superfluid, we shall see that matters are more complicated, in that vortices may appear in the rotating bucket. As a result, one does not encounter the dramatic manifestation of superfluid behavior predicted for a homogeneous superfluid. Nevertheless, the above fictitious experiment is conceptually very important, as it illustrates an essential feature of superfluids, namely their rigidity against transverse probes. The physical importance of such a concept is even more obvious in superconductors, since in that case it corresponds to the Meissner effect.

2. Superfluid flow is resistance free: when a superfluid flows in a thin capillary, it is not slowed down by the walls. The flow thus persists without any pressure head applied to the end of the capillary.

Such a property is surely spectacular. However, although more conspicuous than rigidity against transverse probes, it is conceptually far more subtle, and in a sense less fundamental. In the course of Chapters 4–6, it will become clear that the neatest way of characterizing superfluidity is in terms of irrotational flow, rather than by the absence of resistance.

The two properties of irrotational flow and resistance-free flow constitute the key to superfluid behavior. The preceding qualitative discussion is expanded into a detailed theory in the present chapter for irrotational flow, and in Chapter 5 for resistance-free flow. However, before going into such a theory, it is perhaps useful to set forth an outline of what we wish to do.

Clearly, the formalism of Chapters 2 and 3 is inadequate to treat superfluid flow, since it was only concerned with equilibrium situations in which the condensate is at rest. What we need is a description of superfluid motion, and of the way in which it is set up. For that purpose we can adopt either of two attitudes.

The more general approach is to construct wave functions which explicitly display *condensate motion.* In such wave-functions, the particles are no longer condensed in the state $k = 0$, but in a state with a non-uniform wave-function describing a non-zero velocity of the condensate. The superfluid flow is thus characterized by a change in some suitably defined "condensate wave-function." Once this wave-function is known, one may study the stability of the state under consideration, and look for the best system wave-function in the presence of a given external probe. This point of view will be adopted in Chapters 5 and 10.

The alternative approach is to consider a system at rest, and to describe its response to an external probe as a virtual creation of elementary excitations in the system in its ground state. For instance, in the rotating bucket experiment, we may express the effect of the Coriolis force as creating phonons and rotons in a superfluid at rest. The resulting current, if any, is then viewed as a quasi-particle current. Such an approach, which is closely related to the methods used in Chapters 2 and 3, will be adopted in the present chapter as a description of the response to a transverse probe.

It must, however, be realized that the response function method is less general than that of explicit wave-function construction. By starting with the unperturbed wave-functions, one misses a large class of wave-functions, which cannot be obtained by treating the probe as a small perturbation. More precisely, we shall see in Chapter 5 that there exist two types of superfluid flow, which can be described respectively as "longitudinal" flow and "steady" flow, both characterized by a suitably defined "superfluid velocity" field $\mathbf{v}_s(r)$.

In longitudinal flow, one has div $\mathbf{v}_s \neq 0$. Quite generally, the superfluid density ρ_s and velocity $\mathbf{v}_s$ satisfy the continuity equation

$$\frac{\partial \rho_s}{\partial t} + \text{div } (\rho_s \mathbf{v}_s) = 0 \tag{4.1}$$

(4.1) clearly shows that longitudinal flow is necessarily associated with a time-dependent density fluctuation. We expect such density fluctuations to be closely related to the excitation spectrum discussed in Chapters 2 and 3. We shall indeed show in Chapter 5 that longitudinal flow of the superfluid may be viewed as a virtual phonon emission from the ground state $|0\rangle$. That particular type of motion is thus properly described by a response function formalism. It is only a matter of words whether one prefers to speak of "phonon emission from a fixed condensate" or "weak longitudinal motion of the superfluid." This duality resolves the apparent paradox of the perturbation method, in which superfluid motion is set up without any apparent modification of the condensate.

Conversely, steady superfluid flow involves a velocity field, which is divergence-free (div $\mathbf{v}_s = 0$); the simplest example of such a flow is the vortex structure studied in Chapter 8. Such "hydrodynamic" flow is beyond the range of perturbation theory. The condensate has a really different structure, which must be accounted for from the very beginning of the calculation. In such cases, one must abandon the response function method, and rely on a direct description of superfluid flow.

Despite its lesser generality, we shall consider the response function approach to superfluidity first. We feel that it is conceptually simpler than the detailed description of condensate motion. Moreover, it provides a natural link between the equilibrium treatment of Chapters 2 and 3 and the more general theory of Chapter 5. We shall first analyze in some detail the rotating bucket experiment, and show why it corresponds to a transverse probe acting on the fluid. We shall then calculate the system response and compare it with that to a longitudinal probe. Finally, we shall discuss briefly the "scale" on which the superfluid behavior is observable, thereby introducing the important concept of coherence length.

4.1 Rotating Bucket Experiment: The Transverse Response Function

Let us assume the Bose liquid is put in a bucket which rotates around the z-axis with a constant angular velocity $\boldsymbol{\omega}$ [Blatt, Butler, and Schafroth (1955)]. Let H be the Hamiltonian in the condensate frame of reference, H' that in the frame of reference rotating with the bucket. H' and H are related by the usual equation of classical dynamics

$$H' = H - \boldsymbol{\omega} \cdot \mathbf{L} \tag{4.2}$$

where $\mathbf{L}$ is the total angular momentum

$$\mathbf{L} = \sum_i \mathbf{r}_i \times \mathbf{p}_i \tag{4.3}$$

As usual, the equilibrium situation is obtained when the energy is minimum in a frame of reference in which the walls of the container are at rest (it is only in such a frame that the walls cannot provide energy to the system). The ground state of the rotating bucket is thus obtained by minimizing H', not H. If the angular velocity $\boldsymbol{\omega}$ is small enough, we may consider the term $\boldsymbol{\omega} \cdot \mathbf{L}$ in (4.2) as a perturbation. The rotation of the bucket then acts as a *transverse* probe acting on the liquid; its physical origin is the Coriolis force in the rotating frame of reference.

At $T = 0$, the equilibrium state in the absence of rotation ($\boldsymbol{\omega} = 0$) is the ground state $|0\rangle$, of the Bose liquid. By treating $(-\boldsymbol{\omega} \cdot \mathbf{L})$ as a perturbation acting on $|0\rangle$, we try to express the rotating ground state in terms of a small admixture of excited states into the state, $|0\rangle$. Tacitly, we thereby assume that the rotation of the bucket serves to excite only a small number of quasi-particles, and that it leaves the condensed phase untouched (in the state $k = 0$). In other words, such a perturbation treatment answers the following question: Given an homogeneous condensed phase at rest, can the rotation of the bucket (i.e., the Coriolis force) excite quasi-particles? The resulting current then appears as a quasi-particle current.

We have remarked that such a method provides only a partial answer to the problem at hand. It ignores the possibility of a superfluid steady flow which would be rendered stable by the rotation of the bucket. We shall show in Chapter 8 that it is energetically favorable to introduce vortices in the rotating bucket (except, perhaps, at very low angular velocities). Such a vortex distribution leads to an *average* flow pattern of the fluid which resembles closely that of an ordinary rotating liquid.

Our perturbation treatment of $-\boldsymbol{\omega} \cdot \mathbf{L}$ may seem rather academic, since it does not provide the actual equilibrium situation in a rotating bucket. It nevertheless has a major conceptual importance, as it provides a specific example of superfluid behavior under well-defined assumptions. In short, the state which we shall describe in this chapter corresponds to a sort of "metastable" equilibrium, one which is stable with respect to microscopic fluctuations, but unstable with respect to vortex formation. With this proviso, we may forget the above difficulties: we proceed with our perturbation method, with a clear understanding of what it means physically.

4.2 The Transverse Response Function

We wish to calculate the induced current density in the condensate frame of reference. Let us denote by primed symbols quantities measured in the bucket frame, unprimed symbols referring to those measured in the condensate frame.

With the aid of the expression (4.2) for the Hamiltonian, one may easily verify that the following relations apply.

$$\mathbf{p}'_i = \mathbf{p}_i \tag{4.4a}$$

$$\mathbf{v}'_i = \frac{\mathbf{p}_i}{m} - \boldsymbol{\omega} \times \mathbf{r}_i \tag{4.4b}$$

$\mathbf{p}_i/m$ measures the "absolute" velocity (measured in the condensate frame), while $\mathbf{v}'_i$ is the "relative" velocity. Equation (4.4b) is the usual addition law for velocities. In the same way, the absolute and relative current densities at point **r** are given by

$$\mathbf{J}(\mathbf{r}) = \frac{1}{2}\sum_i \left[\frac{\mathbf{p}_i'}{m}\delta(\mathbf{r}-\mathbf{r}_i) + \delta(\mathbf{r}-\mathbf{r}_i)\frac{\mathbf{p}_i'}{m}\right] \tag{4.5a}$$

$$\mathbf{J}'(\mathbf{r}) = \frac{1}{2}\sum_i [\mathbf{v}_i'\delta(\mathbf{r}-\mathbf{r}_i) + \delta(\mathbf{r}-\mathbf{r}_i)\mathbf{v}_i'] = J(\mathbf{r}) - N\boldsymbol{\omega}\times\mathbf{r} \tag{4.5b}$$

The perturbation $-(\boldsymbol{\omega}\cdot\mathbf{L})$ may be written in the form

$$-\boldsymbol{\omega}\cdot\mathbf{L} = -\sum_i \mathbf{p}_i\cdot\mathbf{A}(\mathbf{r}_i) \tag{4.6}$$

where the transverse "vector potential" $\mathbf{A}(\mathbf{r})$ is given by

$$\mathbf{A}(\mathbf{r}) = \boldsymbol{\omega}\times\mathbf{r} \tag{4.7}$$

The relationship between the present problem and that of the calculation of the response of a charged system to an electromagnetic field is now obvious. The correspondence is outlined in Table 4.1; it permits a fruitful comparison between the properties of liquid helium and superconductors. The current induced by the perturbation (4.6) may be calculated by the response function methods developed in Chapter 2, Vol. I. Let $\mathbf{A}_\mathbf{q}$ and $\mathbf{J}_\mathbf{q}$ be the Fourier transforms of $\mathbf{A}(\mathbf{r})$ and $\mathbf{J}(\mathbf{r})$. For an isotropic system, $\mathbf{A}_\mathbf{q}$ and $\mathbf{J}_\mathbf{q}$ are parallel, both being perpendicular to **q**; let $\eta_\mathbf{q}$ be a unit vector in their mutual direction. By reference to (I.4.121) and (I.4.127), we may easily verify that the response function of interest for the present problem is the transverse current-current response function, which is defined as

$$\chi_\perp(\mathbf{q},\omega) = \sum_n \left|(\eta_\mathbf{q}\cdot\mathbf{J}_\mathbf{q}^+)_{no}\right|^2 \left\{\frac{1}{\omega-\omega_{no}+i\eta} - \frac{1}{\omega+\omega_{no}+i\eta}\right\} \tag{4.8}$$

TABLE 4.1. ***Correspondence Between Transverse Probes for Liquid Helium and a Charged Particle System***

Vector Potential	—	Drift Velocity
Magnetic Field	—	Angular Velocity
Electromagnetic Force	—	Coriolis Force
"Diamagnetic" Current	—	Drift Current
"Paramagnetic" Current	—	"Absolute" Current
Total Electric Current	—	"Relative" Current

The current induced in the bucket is determined by $\chi_\perp(\mathbf{q},0)$, and is

$$\langle \mathbf{J_q} \rangle = -m\chi_\perp(\mathbf{q},0)\mathbf{A_q} \tag{4.9}$$

In the long wave-length limit, we may apply to $\chi_\perp(\mathbf{q},0)$ arguments similar to those used in Chapter 2 for the dynamic form factor. Because of translational invariance, the matrix element $(\eta_\mathbf{q}\cdot\mathbf{J_q}^+)_{no}$ for multi-particle excitations vanishes when $q \to 0$. Since the corresponding excitation energy ω_{no} is finite, we conclude that multi-particle transitions do not contribute to $\chi_\perp(\mathbf{q},0)$ in the long wave-length limit. On the other hand, the matrix element for excitation of a single phonon, with wave vector $\mathbf{q}$, vanishes for reasons of symmetry: a transverse probe cannot couple the ground state to a purely longitudinal excitation, which has axial symmetry around the direction of $\mathbf{q}$. Hence, we see that

$$\lim_{\mathbf{q}\to 0} \chi_\perp(\mathbf{q},0) = 0 \tag{4.10}$$

The result (4.10) is characteristic of a superfluid system at $T = 0$ and is due, for the Bose liquid, to the scarcity of low energy excitations. [For example, in the normal Fermi liquid, the single pair state with momenta $(\mathbf{p},\mathbf{p}+\mathbf{q})$ has no special symmetry around the direction of $\mathbf{q}$, and hence contributes to $\chi_\perp(\mathbf{q},0)$.]

On comparing (4.10) with (4.9), we see that on a macroscopic scale $(\mathbf{q} \to 0)$, the induced current $\mathbf{J}(\mathbf{r})$ in the fixed frame vanishes. Thus, the liquid does not rotate with the bucket; it remains at rest, tied to the condensed phase. If this were really the case, the effect would be readily observable: on rotating the bucket, one would find that the surface of the liquid remains flat, instead of taking the usual meniscus shape. Actually, such a conclusion is only valid for extremely small angular velocities ω. Under normal experimental conditions, there appear vortices in the superfluid, which act to restore the usual meniscus at the surface.

Our result (4.10) is nonetheless important as it means that a transverse probe cannot set up a current of elementary excitations. In other words, the macroscopic current due to the elementary excitations is *irrotational.* Such a statement is typical of a superfluid system. Its extension to finite temperatures will lead naturally to a definition of ρ_n in the two-fluid model (see Chapter 6).

4.3 Longitudinal vs. Transverse Response: Long Range Order

Although superfluid systems do not respond to macroscopic transverse probes, they do respond to *longitudinal* probes. For instance, a scalar (density) probe, which is equivalent to a time-dependent longitudinal vector probe, induces density fluctuations; the corresponding response is measured by the dynamic form factor $S(\mathbf{q},\omega)$ studied in Chapter 2.

In order to show the difference between longitudinal and transverse response, we consider the fictitious problem of a static longitudinal vector probe. The corresponding Hamiltonian is

$$H - \sum_i \mathbf{p}_i \cdot \mathbf{A}(\mathbf{r}_i) \tag{4.11}$$

where the vector potential $\mathbf{A}(\mathbf{r})$ is now expressed as

$$\mathbf{A}(\mathbf{r}) = \mathbf{grad}\ \phi = \sum_{\mathbf{q}} i\mathbf{q}\phi_{\mathbf{q}} e^{i\mathbf{q}\cdot\mathbf{r}} \tag{4.12}$$

The coupling term in (4.11) is formally similar to the perturbation (4.6), but for the fact that $\mathbf{A}$ is a gradient rather than a curl. The velocity of the ith particle is

$$\mathbf{v}_i = \frac{\mathbf{p}_i}{m} - \mathbf{A}(\mathbf{r}_i) \tag{4.13}$$

from which one can derive the current density

$$\mathbf{J}(\mathbf{r}) = \frac{1}{2}\sum_i [\mathbf{v}_i \delta(\mathbf{r}-\mathbf{r}_i) + \delta(\mathbf{r}-\mathbf{r}_i)\mathbf{v}_i] \tag{4.14}$$

[compare with (4.5b)].

Actually, such a static longitudinal probe is unphysical, as it corresponds simply to a gauge transformation. It is easily verified that the perturbed ground state of (4.11) corresponds to the wave-function

$$\exp\left[i\sum_i \phi(r_i)\right]|0\rangle \tag{4.15}$$

where $|0\rangle$ is the ground state of H. On putting (4.15) into the expression for $\mathbf{J}(\mathbf{r})$, we find that the current vanishes. The only effect of our static longitudinal probe is thus to add an arbitrary phase factor to the wave-function of the system.

It is nevertheless interesting to consider in some detail how this gauge invariance comes about. The current $\mathbf{J}(\mathbf{r})$ is a sum of two terms, namely the gauge current

$$-\mathbf{NA}(\mathbf{r}) \tag{4.16}$$

(arising from the second term in $\mathbf{v}_i$), plus the term

$$\langle \mathbf{J} \rangle = \left\langle \sum_i \frac{\mathbf{p}_i}{m} \right\rangle \tag{4.17}$$

arising from the distortion of the ground state wave-function by the longitudinal probe. The two terms must cancel one another in order to ensure gauge invariance. Let us Fourier-analyze in space the probe and the system response. The induced current, $\langle \mathbf{J}_\mathbf{q} \rangle$, may be written as

$$\langle \mathbf{J}_\mathbf{q} \rangle = -m\chi_\parallel(\mathbf{q}, 0)\mathbf{A}_\mathbf{q} \tag{4.18}$$

where $\chi_\parallel(\mathbf{q}, \omega)$ is the longitudinal current-current response function, equal to

$$\chi_\parallel(\mathbf{q}, \omega) = \sum_n \frac{(\mathbf{q} \cdot \mathbf{J}_q^+)^2_{\text{no}}}{q^2} \left\{ \frac{1}{\omega - \omega_{\text{no}} + i\eta} - \frac{1}{\omega + \omega_{\text{no}} + i\eta} \right\} \tag{4.19}$$

On making use of current conservation, (I.2.27), and the f-sum rule (I.2.29), we see that

$$\chi_\parallel(\mathbf{q}, 0) = -\frac{N}{m} \tag{4.20}$$

which ensures the cancellation of $\langle \mathbf{J} \rangle$ with the gauge current.

We can reproduce for $\chi_\parallel$ the arguments used for $\chi_\perp$. Again, multiparticle excitations do not contribute to $\chi_\parallel(\mathbf{q}, 0)$ in the limit $\mathbf{q} \rightarrow 0$; on the other hand, the excitation of a single phonon is no longer forbidden for reasons of symmetry: it will completely determine $\chi_\parallel(\mathbf{q}, 0)$ in this limit. Put another way, the only effect of a macroscopic static longitudinal probe will be to add to the ground state a certain amount of long wave-length phonons. This conclusion is indeed obvious if we look at the exact wave-function (4.15). For a weak probe, the function is small, so that we can expand the exponential. The perturbed wave-function thus becomes

$$\left[1 + i \sum_i \phi(\mathbf{r}) \right] |0\rangle = \left(1 + i \sum_\mathbf{q} \phi_\mathbf{q} \rho_\mathbf{q}^+ \right) |0\rangle \tag{4.21}$$

To the extent that $\phi(\mathbf{r})$ is slowly varying, (4.21) clearly represents the ground state $|0\rangle$ plus a certain admixture of long wave-length phonons, whose wave-function is $\rho_\mathbf{q}^+ |0\rangle$ [see (3.6)].

On comparing (4.20) with (4.10), we see that for a superfluid system when $\mathbf{q} \to 0$ the limits of $\chi_\perp(\mathbf{q}, 0)$ and $\chi_\parallel(\mathbf{q}, 0)$ are different. At first sight, such a result is rather surprising, as one cannot, after all, distinguish between a longitudinal and a transverse probe in this limit. For an isotropic system, one would thus expect

$$\lim_{\mathbf{q}\to 0} \chi_\perp(\mathbf{q}, 0) = \lim_{\mathbf{q}\to 0} \chi_\parallel(\mathbf{q}, 0) = -\frac{N}{m} \tag{4.22}$$

The fact that (4.22) does not apply to superfluid systems constitutes a direct proof of the existence of *long range order* in the fluid.

In order to show this[1] let us write the general response function $\chi_{\mu\nu}(\mathbf{q}, 0)$ as a Fourier transform

$$\chi_{\mu\nu}(\mathbf{q}, 0) = \int d^3\mathbf{r}\, \chi_{\mu\nu}(\mathbf{r}) e^{-i\mathbf{q}\cdot\mathbf{r}} \tag{4.23}$$

$\chi_{\mu\nu}(\mathbf{r})$ gives the current along the μ axis at a point $\mathbf{r}$ set up by a static point vector perturbation (in the ν direction) at the origin. It may be written as

$$\chi_{\mu\nu}(\mathbf{r}) = \int_0^\infty dt\, [J_\mu(\mathbf{r}, t), J_\nu(0, 0)] \tag{4.24}$$

If $\chi_{\mu\nu}(\mathbf{r})$ has a finite range, we may expand the exponential $e^{i\mathbf{q}\cdot\mathbf{r}}$ in (4.23), and keep only the leading term. The direction of $\mathbf{q}$ then disappears from the identity: we are thus led to the relation (4.22). The fact that in a Bose liquid (4.22) is not satisfied proves that in such a system the correlation function $\chi_{\mu\nu}(\mathbf{r})$ has an infinite range. Such a property indicates the presence of long range order, arising from the coherence of the condensed phase.

In a normal Fermi liquid, by contrast, the distortion induced by a static point perturbation has a finite range (such a conclusion is not valid for a finite frequency perturbation, as in that case the distortion may propagate away from the source). We then expect $\chi_{\mu\nu}(\mathbf{r})$ to have a finite range of atomic size, so that (4.22) should hold: such an expectation is substantiated by a detailed calculation based on the Landau theory. In our rotating bucket experiment, this means that the absolute velocity, given by (4.9), is equal to the drift velocity: the liquid is at rest in the rotating frame of reference. A normal Fermi liquid is thus dragged along by the bucket, as one would expect in the absence of any peculiar superfluid behavior. (The corresponding current is produced by an adequate choice of elementary excitations, created by the Coriolis force.)

[1]This proof was suggested to us by P. C. Martin.

The result (4.10) therefore represents a very convenient criterion for superfluidity. It offers a simple way of distinguishing between "normal" fluids (with short range order) and superfluids (possessing long range order); its generalization to finite temperatures will be given in Chapter 6.

4.4 Coherence Length

Thus far we have only considered the current response functions in the limit $\mathbf{q} \to 0$. When $\mathbf{q}$ is finite, the longitudinal response function $\chi_{\parallel}(\mathbf{q}, 0)$, continues to be given by its long wave-length value, (4.20). On the contrary, the transverse part $\chi_{\perp}(\mathbf{q}, 0)$ no longer vanishes, since the multiparticle excitations contribute a term proportional to some power of $\mathbf{q}$.

For sufficiently large values of $\mathbf{q}$, it is straightforward to calculate $\chi_{\perp}(\mathbf{q}, 0)$, because in this limit particle interactions play a negligible role. One finds

$$\lim_{\mathbf{q} \to \infty} \chi_{\perp}(\mathbf{q}, 0) = -\frac{N}{m} \tag{4.25}$$

A schematic plot of $\chi_{\perp}(\mathbf{q}, 0)$ is shown in Fig. 4.1; $\chi_{\perp}$ displays a maximum at some value $q = q_c$. The length q_c^{-1} acts as a characteristic length of the problem: it marks the minimum distance over which one can observe *perfect* superfluid order.

When $q \sim q_c$, the response function $\chi_{\perp}$ is q-dependent; the relation between the vector potential $A(\mathbf{r})$ and the current $J(\mathbf{r}')$ is then non-local. The characteristic length q_c^{-1} measures the width of the region in which the potential $A(r)$ influences the current at a given point $\mathbf{r}'$. From this point of view, q_c^{-1} is the range of the current response function.

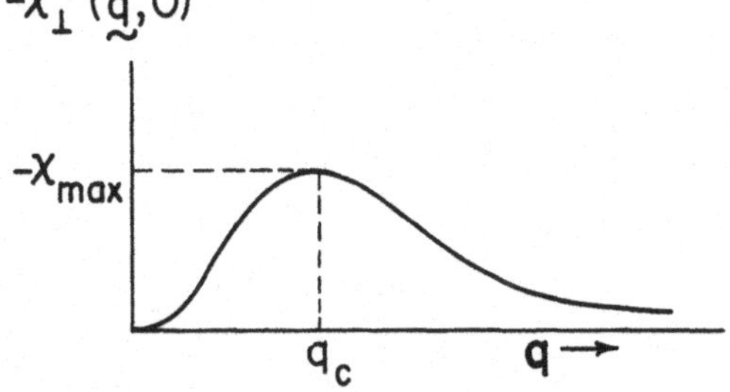

FIGURE 4.1. *Schematic behavior of* $\chi_{\perp}(\mathbf{q}, 0)$.

The general question of characteristic lengths will be discussed in Chapter 10. We shall see that $\mathbf{q}_c$ is an increasing function of the interaction strength between particles. Similarly, it may be shown that $\chi_{\max}$ increases with particle interactions. For a free Bose gas, $\chi_{\perp}(q) = 0$ for all q; for the weakly interacting Bose gas discussed in Chapter 9, one finds that

$$-\chi_{\max} \ll \frac{N}{m} \tag{4.26}$$

In such a case, the response to a transverse probe remains very small for any q. In actual liquid ^{4}He, the coupling is of intermediate strength: $-\chi_{\max}$ is comparable to N/m, and $\mathbf{q}_c$ is of the order of the roton wave vector.

CHAPTER 5

SUPERFLUID FLOW: MACROSCOPIC LIMIT

In the preceding chapters, we have been mostly concerned with the ground state, $|0\rangle$, of a Bose liquid; we have discussed at length its structure, its excitation spectrum, and its response to a weak external probe. We have seen that all these properties were deeply influenced by the existence of a *condensate*, consisting in the n_o particles which occupy the single state $\mathbf{k} = 0$. Such a condensate is clearly at rest; it possesses a uniform density and carries no current.

Actually, the fundamental feature of Bose condensation is the *macroscopic* occupation of a *single* quantum state, whatever that state may be. As long as such a macroscopically-occupied state exists, the system will exhibit superfluid behavior. According to the general principles of quantum mechanics, we may associate to this generalized condensate a wave-function, ϕ. That wave-function will be a constant in the case we have thus far considered, that of macroscopic occupation of the $\mathbf{k} = 0$ state. More generally one expects to find a condensate wave-function which varies with space and time, $\phi(\mathbf{r}, t)$. Such a condensate may possess density fluctuations and carry current. We have essentially introduced a new "degree of freedom," namely that of *superfluid motion.* Such a motion arises as a consequence of the condensate structure, and involves a motion of the condensed particles *as a whole.* Superfluid flow thus appears as a "collective" phenomenon, in which the particles move together in order to preserve the macroscopic occupation of a single state.

Once a steady superfluid flow is established, through the appropriate condensate motion, the flow will be unusually stable. Consider, as an example, the flow of a superfluid in a pipe. It may be altered by the pipe walls in only two ways:

1. By a modification in the condensate wave-function.
2. By a depletion of the condensate as a result of the creation of elementary excitations.

Modification of the condensate as a whole is certainly difficult to bring about, since it involves *simultaneous* action on a *macroscopic* number of particles. We may therefore rule it out as a possible source of resistance to superfluid flow. The second possibility, that the obstacle might act to create elementary excitations, costs energy, which must, in fact, be supplied by the fluid flow itself. We shall see that below a certain critical velocity, such energy is not available.

Under suitable conditions, therefore, the system will be in a state of *metastable* equilibrium, in which flow occurs without viscous damping. The equilibrium is metastable, because obviously there exists a state of lower energy, that of the fluid at rest; the essential point is that no transitions to the latter state will take place. Metastable superfluid flow is only possible because it involves condensate motion; it is in this way that it appears as a direct consequence of Bose condensation.

The present chapter is devoted to the detailed description of condensate motion. We begin by writing down a condensate wave-function; we are led thereby to a precise definition of the "superfluid velocity" $\mathbf{v}_s$. We use this to establish two fundamental properties of superfluid flow: the irrotational character of the flow and the quantization of circulation. The physical origin of superfluid flow is then discussed in detail, as is its relation to Bose condensation. In the latter part of the chapter, we consider anew the response of the condensate to an external longitudinal "density" probe. This response, which was considered in Chapter 2 as phonon emission from a fixed condensate, will here be described as a longitudinal motion of the condensate. We thus establish the connections between the two approaches to superfluid behavior.

Throughout this chapter, we shall consider situations in which the flow pattern varies slowly over a coherence length; locally, the superfluid flow can then be considered as a uniform translation. The more difficult problem of rapidly varying velocity fields is considered in Chapter 10.

5.1 Description of Condensate Motion

The simplest example of condensate motion is that of a *uniform translation.* As an example of such a translation, let us consider a non-interacting Bose gas in which all particles are moving at the *same* velocity[1]

$$\mathbf{v}_s = \frac{\hbar \mathbf{q}}{m} \tag{5.1}$$

The corresponding wave-function, expressed in configuration space, is

$$\psi = \prod_i e^{i\mathbf{q}\cdot\mathbf{r}_i} = \exp\left[i\sum_i \mathbf{q}\cdot\mathbf{r}_i\right] \tag{5.2}$$

We note that (5.2) is obtained from the ground state $\psi_o = 1$ by a simple Galilean transformation. Put another way, ψ describes the ground state ψ_o as viewed from a frame of reference moving at velocity $-\mathbf{v}_s$. Such a transformation is equivalent to adding the same amount $\mathbf{q}$ to the momenta of all particles. Thus, in the state ψ described by (5.2), *all* particles are in the *same* quantum state, that with wave vector $\mathbf{q}$. We thus preserve the important feature of *Bose condensation* of all the system particles into a single quantum state. The condensate moves at velocity $\mathbf{v}_s$, in contrast with the case of the ground state ψ_o. Such "condensate motion" may be described as a *coherent* displacement of all condensed particles.

The wave-function (5.2) is a product of factors $e^{i\mathbf{q}\cdot\mathbf{r}_i}$, one for each particle. The quantity

$$\phi(\mathbf{r}) = e^{i\mathbf{q}\cdot\mathbf{r}} \tag{5.3}$$

thus appears as the wave-function of a "condensed" particle, which for brevity we shall describe as the *condensate wave-function* (here a plane wave with vector $\mathbf{q}$). This concept, which here appears to be trivial, forms the keystone of the theory of condensate motion.

Let us now consider an interacting Bose liquid. Since the particle interactions are invariant under translations, we may construct an eigenstate of the system by applying a Galilean transformation to the *exact* ground state wave-function ψ_o. We thus obtain a state characterized by the wave-function

$$\psi = \exp\left[i\sum_i \mathbf{q}\cdot\mathbf{r}_i\right]\psi_o \tag{5.4}$$

[1]In order to display clearly the effects of macroscopic quantization, we retain all factors of $\hbar$ in this chapter.

Again, (5.4) describes the ground state as viewed from a frame of reference moving at a constant velocity $-\hbar\mathbf{q}/m$. It thus corresponds to a *uniform translation* of the Bose liquid. We note that (5.4) is a rigorous eigenstate of the interacting liquid; it may be obtained from (5.2) by adiabatically turning on the interaction between particles.

As a result of the Galilean transformation, all particles are shifted in momentum space by the same amount $\hbar\mathbf{q}$. For instance, the n_o particles which in the ground state were condensed in the state with zero momentum are now condensed in the state with momentum $\hbar\mathbf{q}$. There still exists a condensate, but the condensate now moves at a constant velocity $\hbar\mathbf{q}/m$.

The Galilean transformation also affects the $(n - n_o)$ particles which in the ground state were virtually excited into states with non-zero momentum. Their distribution $n_\mathbf{p}$ is rigidly shifted in momentum space by the same amount $\hbar\mathbf{q}$ as the condensate (as shown in Fig. 5.1). Thus, although these particles are not condensed, they are "tied" to the condensate; they are dragged along when the latter is set in motion. For instance, the moving state described by (5.4) corresponds to a total mass current.

$$\mathbf{J} = n\hbar\mathbf{q} = nm\mathbf{v}_s \tag{5.5}$$

The fact that $\mathbf{J}$ involves n rather than n_o shows clearly that the virtually excited particles take part in superfluid motion to the same extent as the condensate.

The moving state described by (5.4) is easily interpreted in the language of the two-fluid model. If we recall that at $T = 0$ we have $\rho_s = \rho$

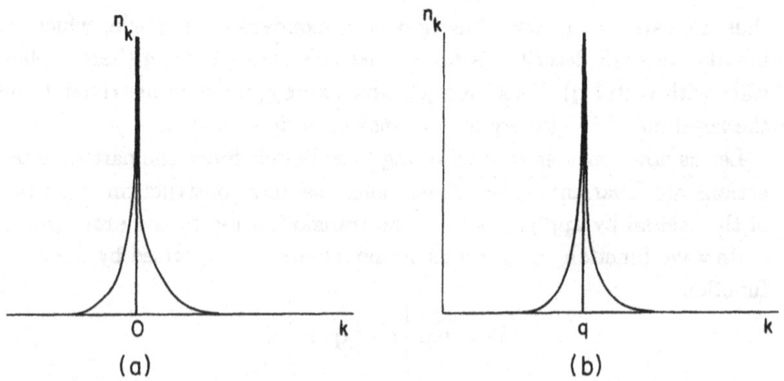

FIGURE 5.1. *Distribution of bare particles: (a) in the ground state, (b) for a state in uniform translation.*

and $\rho_n = 0$, we see that $\mathbf{v}_s$, as given by (5.5), may be identified with the *superfluid velocity* introduced in Chapter 1. This essential physical concept thus acquires a precise definition: it is related to the condensate momentum by equation (5.1).

Let us return to the wave-function (5.4) describing the moving fluid. We may write it as:

$$\psi = \prod_i e^{i\mathbf{q}\cdot\mathbf{r}_i}\psi_o \tag{5.6}$$

Again, $e^{i\mathbf{q}\cdot\mathbf{r}}$ appears as the *condensate wave-function*, which serves to characterize the condensate structure as well as its motion. There are n such factors; we see once more that *all* particles participate in the superfluid motion, even though only a fraction n_o of them are actually condensed in a single quantum state.

The factor ψ_o in (5.6) incorporates all the subtle correlations arising from the interaction between particles. The use of (5.6) implies that these correlations are unaffected by the fluid flow. Such an exact decoupling of the "internal" degrees of freedom (as described by ψ_o) from the main fluid flow [described by the exponentials in (5.6)] is a special feature of uniform translation; it does not persist in the case of non-uniform flow patterns.

The preceding discussion provides an accurate description of uniform translation of the Bose liquid. However, it must be realized that it depends on a specific assumption concerning the nature of such a flow. The choice of (5.4) as a wave-function is essentially arbitrary; it is only dictated by the simplicity of Galilean transformations. Actually, given a condensate with a non-zero momentum $\hbar\mathbf{q}$, there exist a whole series of eigenstates, similar to those found for $\mathbf{q} = 0$. The lowest of these states may be described as a "moving" ground state, which corresponds to a metastable equilibrium (leading to superfluid flow) at zero-temperature. Higher states arise from a corresponding set of "elementary excitations of the moving fluid." The condensate momentum $\hbar\mathbf{q}$ thus appears as a *parameter* of the system wave-function, describing a new degree of freedom. When using (5.4) to describe condensate motion we implicitly assume that the wave-function provides the lowest possible energy for a given $\mathbf{q}$. In fact, we shall show later in this chapter that such an assumption is valid for moderate values of the condensate momentum $\mathbf{q}$. We shall thereby explain the stability of superfluid flow.

We now consider the extension of these considerations to an arbitrary motion of the superfluid condensate. Again, we consider first the non-interacting Bose gas. For this system a wave-function which describes all particles condensed into the same state, $\phi(\mathbf{r})$, will take the form,

$$\psi(\mathbf{r}_1 \ldots \mathbf{r}_n) = \prod_{i=1}^{n} \phi(\mathbf{r}_i) \tag{5.7}$$

where the "condensate" wave-function $\phi(\mathbf{r})$ is assumed to be normalized. The wave-function (5.7) corresponds to a particle density

$$\rho(\mathbf{r}) = N\phi^*(\mathbf{r})\phi(\mathbf{r}) \tag{5.8}$$

and to a current density

$$\mathbf{J}(\mathbf{r}) = -\frac{i\hbar}{2m}N\left[\phi^*(\mathbf{r})\nabla\phi(\mathbf{r}) - \phi(\mathbf{r})\nabla\phi^*(\mathbf{r})\right] \tag{5.9}$$

It is convenient to write $\phi(\mathbf{r})$ in the form

$$\phi(\mathbf{r}) = f(\mathbf{r})e^{iS(\mathbf{r})} \tag{5.10}$$

where f and S are respectively the modulus and the argument of ϕ. It is easily verified that

$$\rho(\mathbf{r}) = Nf^2(\mathbf{r}) \tag{5.11}$$

$$J(\mathbf{r}) = \rho(\mathbf{r})\frac{\hbar}{m}\,\mathbf{grad}\;S(\mathbf{r}) \tag{5.12}$$

The wave-function (5.7) thus describes a *moving* condensate, whose density fluctuations are controlled by the modulus of ϕ; the phase of ϕ determines the fluid velocity $\mathbf{J}/\rho$. In the case of fluid motion which does not involve a density fluctuation ($f = 1$), (5.7) becomes simply

$$\psi(\mathbf{r}_1 \ldots \mathbf{r}_n) = \exp\left[i\sum_{i=1}^{N} S(\mathbf{r}_i)\right] \tag{5.13}$$

The velocity of condensate motion is spatially varying, save in the special case of uniform translation, for which $S(\mathbf{r}) = \mathbf{q}\cdot\mathbf{r}$.

We consider now the extension of such a description to the interacting Bose liquid, specializing first to the case in which there are no density fluctuations. In this case, London (1954), Feynman (1955), and Onsager

(1949) have suggested that condensate motion may be described by the following wave-function:

$$\psi(\mathbf{r}_1 \ldots \mathbf{r}_n) = \exp\left[i \sum_i S(\mathbf{r}_i)\right] \psi_o(\mathbf{r}_1 \ldots \mathbf{r}_n) \tag{5.14}$$

where ψ_o is the exact ground state wave-function. As was the case for uniform translational motion, we may try to pass from (5.13) to (5.14) by adiabatically turning on the interaction between particles. However, such a procedure, which in principle is rigorous, does not yield exactly (5.14), in contrast with the case of uniform translation. Consequently, (5.14) does *not* describe an exact eigenstate. The physical origin of this difficulty is clear: if the condensate varies sufficiently rapidly in space, we may expect that the comparatively short-range interaction between the particles will be altered, so that the wave-function multiplying (5.13) will differ from the ground state wave-function.

Put another way, in (5.14) the superfluid motion is completely decoupled from the short-range correlations which manifest themselves in ψ_o. In practice, one would not expect such a decoupling to be perfect; the London wave-function (5.14) is therefore only an approximate description of superfluid motion, and we must inquire as to the conditions which limit its applicability.

For a uniform translation of the condensate, (5.13) is exact, since it agrees with the wave-function (5.4) obtained by means of Galilean invariance arguments. Next let us suppose that $\mathbf{grad}\, S(\mathbf{r})$ varies only slowly over some coherence length. Viewed from the "atomic" scale, the corresponding condensate motion appears to be an essentially uniform translation. Since the short-range correlations are unaffected by such a translation, we would expect that the wave-function (5.14) furnishes a *locally* correct description of superfluid behavior. In the opposite limit, when S varies appreciably over a coherence length, we expect (5.14) not to be valid.

The above "coherence length" provides a measure of the condensate rigidity. It corresponds to the minimum scale over which one can deform the flow pattern without affecting the microscopic structure of the wave-function. A priori, such a coherence length need not be the same as that introduced earlier (in Chapter 4), which corresponded to the distance at which the transverse current-current response function began to exhibit non-local behavior. The latter characteristic length offers a measure of the point at which the ground state wave-function is no longer rigid in its response to a transverse external disturbance. To the extent that a rapid variation in S may be viewed as a transverse disturbance in its action on the ground state wave-function, the two lengths may be expected to agree. In the case of liquid helium, such is the case to a good degree of approximation; both lengths are of the order of the inter-particle spacing. On the other hand, the lengths may be quite different in the case of impure superconductors.

We next consider the wave-function which describes density fluctuations in a Bose liquid. We pursue once more the analogy with the non-interacting Bose gas, and take for this wave-function

$$\psi(\mathbf{r}_1 \ldots \mathbf{r}_n) = \prod_{i=1}^{n} \phi(\mathbf{r}_i)\, \psi_o(\mathbf{r}_1 \ldots \mathbf{r}_n) \tag{5.15}$$

where $\phi(\mathbf{r})$ now has a varying modulus:

$$\phi(\mathbf{r}) = f(\mathbf{r}) e^{iS(\mathbf{r})} \tag{5.16}$$

[compare with (5.7) and (5.14)]. $\phi(\mathbf{r})$, which describes the condensate structure, may be considered as the *condensate wave-function*. In this chapter, we shall assume $\phi(\mathbf{r})$ to be normalized to 1 (a different convention will be adopted in Chapter 10).

In practice, the use of (5.15) is much more open to criticism than that of (5.14), as the particle correlations are likely to be much affected by the fluctuations of density. A wave-function such as (5.15) only works for a *small amplitude, long wave length* density fluctuation, for which we may write

$$f(\mathbf{r}) = 1 + \alpha(\mathbf{r}) \tag{5.17}$$

where α is small. We may then expand the product of factors f up to first order in α:

$$\prod_{i=1}^{N} [1 + \alpha(\mathbf{r}_i)] \cong 1 + \sum_{i=1}^{N} \alpha(\mathbf{r}_i)$$

On Fourier-expanding the function $\alpha(\mathbf{r})$, we write (5.15) in the form

$$\psi(\mathbf{r}_1 \ldots \mathbf{r}_n) = \left(1 + \sum_{\mathbf{q}} \alpha_{\mathbf{q}} \rho_{\mathbf{q}}^{+}\right) \exp\left[i \sum_{i} S(\mathbf{r}_i)\right] \psi_o(\mathbf{r}_1 \ldots \mathbf{r}_n) \tag{5.18}$$

When $S = 0$, (5.18) describes a small admixture of long wave length phonons into the ground state ψ_o, and thus corresponds to a small density fluctuation. When $S \neq 0$, this density fluctuation is superimposed on the current flow described by S.

The above discussion enables us to appreciate the extent to which a generalized London wave-function

$$\psi(\mathbf{r}_1 \ldots \mathbf{r}_n) = \prod_i \left[f(\mathbf{r}_i)\, e^{iS(\mathbf{r}_i)} \right] \psi_o(\mathbf{r}_1 \ldots \mathbf{r}_n) \tag{5.19}$$

may be used to describe arbitrary superfluid motion. The wave-function (5.19) offers an adequate description provided:

1. f and S vary slowly over a coherence length.
2. $f \cong 1$.

Throughout the remainder of this chapter, we assume those conditions are met; we therefore will consider only macroscopic motion which involves small amplitude density fluctuations. The more general case of motion on a *microscopic* scale will be considered in Chapter 10.

5.2 Density, Current, and Energy Associated with Condensate Motion

The wave-function (5.19) may be used to calculate some of the macroscopic properties of the liquid, for instance its density, current density, and energy. The density at point r is equal to

$$\rho(\mathbf{r}) = \int d^3\mathbf{r}_1 \ldots d^3\mathbf{r}_n\, |\psi(\mathbf{r}_1 \ldots \mathbf{r}_n)|^2 \sum_i \delta(\mathbf{r} - \mathbf{r}_i) \tag{5.20}$$

Similarly, the current density at point $\mathbf{r}$, $\mathbf{J}(\mathbf{r})$, may be written as

$$\mathbf{J}(\mathbf{r}) = \int d^3\mathbf{r}_1 \ldots d^3\mathbf{r}_n \psi(\mathbf{r}_1 \ldots \mathbf{r}_n)\, \mathbf{J}_{\text{op}}(\mathbf{r}) \psi(\mathbf{r}_1 \ldots \mathbf{r}_n) \tag{5.21}$$

where the current density operator, $\mathbf{J}_{\text{op}}(\mathbf{r})$, is defined as

$$\mathbf{J}_{\text{op}}(\mathbf{r}) = \frac{1}{2m} \sum_i \left[\delta(\mathbf{r} - \mathbf{r}_i)(-i\hbar\nabla_i) + (-i\hbar\nabla_i)\,\delta(\mathbf{r} - \mathbf{r}_i) \right] \tag{5.22}$$

Let us replace the wave-function ψ by its expression (5.19). After some algebra, we write $\mathbf{J}(\mathbf{r})$ in the following form

$$\mathbf{J}(\mathbf{r}) = \int d^3\mathbf{r}_1 \ldots d^3\mathbf{r}_n \prod_j |f(\mathbf{r}_j)|^2 \times \sum_i \delta(\mathbf{r} - \mathbf{r}_i) \left\{ \frac{\hbar}{m} \nabla S(\mathbf{r}_i)\, \psi_o^2 - \frac{i\hbar}{2m} \left[\psi_o^* \nabla_i \psi_o - \psi_o \nabla_i \psi_o^* \right] \right\} \tag{5.23}$$

The last term in (5.23) corresponds to the current density in the ground state, and vanishes by symmetry (the wave-function ψ_o is in fact real). Thus, the only contribution to $\mathbf{J}(\mathbf{r})$ arises from the first term in the brackets. On making use of (5.20), we find

$$\mathbf{J}(\mathbf{r}) = \frac{\hbar}{m}\nabla S(\mathbf{r}) \int d^3\mathbf{r}_1 \ldots d^3\mathbf{r}_n \sum_i \delta(\mathbf{r} - \mathbf{r}_i)\,|\psi|^2$$

$$= \rho(\mathbf{r})\frac{\hbar}{m}\nabla S(\mathbf{r}) \tag{5.24}$$

The current is superfluid in character; locally, all the particles move together at the same velocity, dragged by the condensate motion.

The superfluid velocity $\mathbf{v}_s$ at point $\mathbf{r}$ is naturally defined by the relation

$$\mathbf{J}(\mathbf{r}) = \rho(\mathbf{r})\mathbf{v}_s(\mathbf{r}) \tag{5.25}$$

On comparing (5.25) with (5.24), we see that

$$\mathbf{v}_s = \frac{\hbar}{m}\,\mathbf{grad}\, S \tag{5.26}$$

The superfluid velocity is thus proportional to the gradient of the *phase* of the condensate wave-function.

Let us return to the density, given by (5.20). It is convenient to calculate its Fourier transform

$$\langle\rho_\mathbf{q}^+\rangle = \langle\psi|\rho_\mathbf{q}^+|\psi\rangle = \int d^3\mathbf{r}_1 \ldots d^3\mathbf{r}_n\,|\psi|^2 \sum_i e^{+i\mathbf{q}\cdot\mathbf{r}_i} \tag{5.27}$$

On replacing ψ by its expression (5.18), we see that the phase factor, e^{iS}, does not contribute to the density fluctuations. Up to first order in α, we may write

$$\langle\rho_\mathbf{q}^+\rangle = 2\alpha_q \langle\psi_o\,|(\rho_\mathbf{q}\rho_\mathbf{q}^+)|\psi_o\,\rangle = 2\alpha_\mathbf{q} N S_\mathbf{q} \tag{5.28}$$

where $S_\mathbf{q}$ is the static form factor. For a non-interacting Bose gas, $S_\mathbf{q} = 1$: in that case, we find

$$\langle\rho_\mathbf{q}\rangle = 2N\alpha_\mathbf{q} \tag{5.29a}$$

from which it follows that

$$\rho(\mathbf{r}) = 2N\alpha(\mathbf{r}) \tag{5.29b}$$

The result (5.29) is not valid for an *interacting* Bose liquid; $S_\mathbf{q}$ is given by (2.29). We thus find

$$\langle\rho_\mathbf{q}\rangle = \frac{N\hbar q}{ms}\alpha_\mathbf{q} \tag{5.30}$$

in place of (5.29a). (We have restored in $S_{\mathbf{q}}$ a missing factor of $\hbar$.) In ordinary space, $\rho(\mathbf{r})$ involves the gradient of $\alpha(\mathbf{r})$, rather than α itself. The effect of particle interactions is therefore extremely important.

Let us finally calculate the energy E of the state described by the wave-function (5.18), as measured from the ground state energy E_o. To the extent that ψ is normalized, the difference $(E - E_o)$ is equal to

$$E - E_o = \int d^3\mathbf{r}_1 \ldots d^3\mathbf{r}_n \, \psi_o^* \prod_i f(\mathbf{r}_i)\, e^{-i\sum_i S(\mathbf{r}_i)} \left[H, \prod_j f(\mathbf{r}_j)\, e^{i\sum_j S(\mathbf{r}_j)}\right]\psi_o \tag{5.31}$$

The calculation of $(E - E_o)$ is thus reduced to an evaluation of the commutator in (5.31). Let us first commute H with the exponential factor $\exp\left[\sum_i S(\mathbf{r}_i)\right]$. On making use of the identity

$$e^{-iS} H e^{iS} = H + i[H, S] + \frac{i^2}{2}[[H, S], S] + \ldots \tag{5.32}$$

we find the following contribution to $(E - E_o)$

$$(E - E_o)_1 = \int d^3\mathbf{r}_1 \ldots d^3\mathbf{r}_n \, |\psi|^2 \sum_i \frac{[\hbar \nabla S(\mathbf{r}_i)]^2}{2m} \tag{5.33}$$

(the detailed calculation is left as an exercise to the reader). On referring to (5.20), we see that (5.33) may be written as

$$(E - E_o)_1 = \int d^3\mathbf{r}\, \rho(\mathbf{r}) \frac{[\hbar \nabla S(\mathbf{r})]^2}{2m} \tag{5.34}$$

If we now use the definition (5.26) of $\mathbf{v}_s$, we obtain

$$(E - E_o)_1 = \int d^3\mathbf{r}\, \rho(\mathbf{r}) \frac{1}{2} m \mathbf{v}_s^2(\mathbf{r}) \tag{5.35}$$

(5.35) is clearly the *kinetic energy* associated with the fluid flow velocity $\mathbf{v}_s(\mathbf{r})$.

It now remains to commute H with the factor $\prod_j f(\mathbf{r}_j)$, which we write in the form (5.18). We have seen that, for the long wave lengths in which we are interested, the operator $\rho_{\mathbf{q}}^+$ creates a phonon with energy $\varepsilon_q = sq$. It follows that

$$\left[H, 1 + \sum_{\mathbf{q}} \alpha_{\mathbf{q}} \rho_{\mathbf{q}}^+\right] = \sum_q \alpha_{\mathbf{q}} \varepsilon_q \rho_{\mathbf{q}}^+ = \sum_{\mathbf{q}} \hbar s q \alpha_{\mathbf{q}} \rho_{\mathbf{q}}^+ \tag{5.36}$$

On substituting (5.36) into (5.31), we find the following contribution to $(E - E_o)$:

$$(E - E_o)_2 = \int d^3\mathbf{r}_1 \ldots d^3\mathbf{r}_n \psi_o^* \left(1 + \sum_{\mathbf{q}} \alpha_{\mathbf{q}} \rho_{\mathbf{q}}\right) \sum_{\mathbf{q}} \hbar s q \alpha_{\mathbf{q}} \rho_{\mathbf{q}}^+ \psi_o \quad (5.37)$$

which may be written in the form

$$(E - E_o)_2 = \sum_{\mathbf{q}} \hbar s q \alpha_{\mathbf{q}}^2 \langle \psi_o | \rho_{\mathbf{q}} \rho_{\mathbf{q}}^+ | \psi_o \rangle = \sum_{\mathbf{q}} N \hbar s q \alpha_{\mathbf{q}}^2 S_{\mathbf{q}} \quad (5.38)$$

On making use of (2.29) and (5.30), we obtain:

$$(E - E_o)_2 = \sum_{\mathbf{q}} \frac{N\hbar^2 q^2 \alpha_{\mathbf{q}}^2}{2m} = \frac{ms^2}{2N} \sum_{\mathbf{q}} \langle \rho_{\mathbf{q}} \rangle^2 \quad (5.39)$$

Let $\delta\rho(r)$ be the fluctuation about the equilibrium density. (5.39) may be written as

$$(E - E_o)_2 = \frac{ms^2}{2N} \int d^3\mathbf{r}\, [\delta\rho(\mathbf{r})]^2 \quad (5.40)$$

(5.40) represents the *potential energy* associated with the density fluctuations. This energy may in fact be obtained by purely macroscopic arguments. According to (3.8), s^2 is related to the compressibility κ of the liquid. (5.40) may thus be written as

$$(E - E_o)_2 = \frac{1}{2N^2\kappa} \int d^3\mathbf{r}\, [\delta\rho(\mathbf{r})]^2 \quad (5.41)$$

which is just what one would obtain by elementary methods.

On collecting (5.34) and (5.39), we may write the total energy $(E - E_o)$ in either of the forms

$$(E - E_o) = \frac{\hbar^2}{2m} \int d^3\mathbf{r} \rho(\mathbf{r}) \left\{ [\nabla S(\mathbf{r})]^2 + [\nabla \alpha(\mathbf{r})]^2 \right\} \quad (5.42)$$

$$= \frac{1}{2} \int d^3\mathbf{r} \left\{ \rho(\mathbf{r}) m \mathbf{v}_s^2(\mathbf{r}) + \frac{ms^2}{N} \delta\rho^2(\mathbf{r}) \right\} \quad (5.43)$$

We note the complete decoupling of the kinetic and potential terms, arising respectively from the phase and modulus of the condensate wavefunction $\phi(\mathbf{r})$.

5.3 Irrotational Character of Superfluid Flow

The superfluid velocity $\mathbf{v}_s$ is given by equation (5.26). It follows that

$$\text{curl } \mathbf{v}_s(\mathbf{r}) = 0 \tag{5.44}$$

We thus arrive once again at the conclusion that the superfluid flow is *irrotational*. In the framework of the present section, this fundamental property is seen to be a direct consequence of the macroscopic occupation of a single quantum state, since the flow velocity is simply proportional to the gradient of the condensate wave-function.

Such a derivation of superfluid flow is complementary to that presented in Chapter 4. There we saw that at zero temperature the system did not respond to a macroscopic transverse probe. Since system response may equally well be viewed as a small superfluid current superimposed on the ground state, such a result means that it is impossible to create a weak transverse superfluid current fluctuation. The result (5.44) represents an extension of this property to arbitrarily large superfluid velocities. Let us emphasize that in both the preceding section and the present one we have made fundamental use of the macroscopic occupation of a single quantum state; the derivation of (5.44) by the methods of the present section is perhaps more direct, while the methods of the preceding section show, in somewhat more perspicuous fashion, the limitations on the validity of the London wave-function, (5.19).

We mentioned at the beginning of Chapter 4 that there exist two types of superfluid flow, namely *longitudinal* and *steady* flow. Longitudinal flow implies the existence of time dependent density fluctuations, and can be viewed as a phonon emission from a fixed condensate; it will be discussed at the end of this chapter. Here we consider *steady* flow, for which the density of the system remains constant. The velocity field $\mathbf{v}_s(\mathbf{r})$ must then satisfy the conservation equation

$$\text{div } \mathbf{v}_s = 0 \tag{5.45}$$

Furthermore, the normal component of $\mathbf{v}_s$ at the surface of the system must vanish. For a singly connected system, the only regular solution of (5.44) and (5.45) satisfying such a boundary condition is $\mathbf{v}_s = 0$. Situations in which $\mathbf{v}_s \neq 0$ are thus only possible for a truly multiply connected system (for example, a ring around which the current flows), or for one which is effectively multiply connected (as a consequence of singularities in the system near which the London wave-function breaks down). An example of the latter case is the vortex line discussed in Chapter 8.

5.4 Quantization of Circulation

The velocity $\mathbf{v}_s$ possesses a fundamental property, which is characteristic of superfluid systems: its circulation along any closed curve Γ,

$$C = \int_\Gamma \mathbf{v}_s \cdot d\mathbf{l} \tag{5.46}$$

is *quantized*: it is equal to an integer multiple of a fundamental quantum h/m. This important property is the counterpart for the Bose liquid of the quantization of magnetic flux in superconductors.

The physical origin of the quantization is best understood by considering a one-dimensional system, of length L, which is closed on itself to form a ring. Since the wave-function must be unaffected by a translation from x to $(x+L)$, any wave vector q can only take discrete values, of the form

$$q = \frac{2\pi}{L} n \tag{5.47}$$

where n is an integer. In order to set up a superfluid current, we must impart to the condensate a non-zero wave vector q. The current carried by the condensate is then equal to

$$J_s = \frac{N\hbar q}{m} \tag{5.48}$$

It follows from (5.47), that J_s is quantized,

$$J_s = n\frac{Nh}{mL} \tag{5.49}$$

Because of the factor N in (5.49), the basic quantum of current is a *macroscopic* rather than a microscopic quantity. Such a macroscopic effect of quantization is a direct consequence of the macroscopic occupation of a single quantum state. If we change the momentum of the latter by one quantum $2\pi/L$, the corresponding change in current is N times larger. In a normal system, we can change the momentum of a *single* particle by an amount $2\pi/L$: instead of (5.49), the quantum of current is then h/mL. Since the latter quantum is vanishingly small, the quantization of current may be ignored on any macroscopic scale.

Let us return to the case of a Bose liquid. The velocity $\mathbf{v}_s$ is equal to J_s/N. The circulation of the velocity around the ring is simply $v_s L$, and is thus of the form

$$C = n\frac{h}{m} \tag{5.50}$$

where n is an integer. We have thus proved, in this particular case, that the circulation of the superfluid velocity $\mathbf{v}_s$ is *quantized* in units h/m.

We now proceed to demonstrate the quantization of circulation by a more general method, one which is not restricted to a one-dimensional system. Let us replace $\mathbf{v}_s$ by its expression (5.26); the circulation C along a closed curve, Γ, may then be written as

$$C = \oint_\Gamma \mathbf{v}_s \cdot d\mathbf{l} = \frac{\hbar}{m} \oint_\Gamma \mathbf{grad}\, S \cdot d\mathbf{l} \tag{5.51}$$

$mC/\hbar$ is thus equal to the change of S when one goes around the curve (which means that S is a multi-valued function). However, the wave-function of the condensate, equal to $e^{iS(\mathbf{r})}$, must be *single valued*. Thus, S can only change by a multiple of 2π when one goes around a closed circuit, from which fact (5.50) follows directly. The quantization of circulation is seen to be a direct consequence of the requirement that the condensate wave-function be single-valued. It thus applies to any distribution of superfluid current.

The quantization of circulation is a major physical feature of superfluid flow. Together with the irrotational character of the flow, equation (5.44), it governs the way in which a superfluid can be set in motion. We shall illustrate in Chapter 8 the way in which these concepts operate, by considering a specific example, that of the vortical motion of a superfluid Bose liquid.

5.5 Flow Without Resistance: Landau Criterion

According to our previous discussion, the wave-function describing a uniform translation of the fluid is obtained by rigidly shifting the ground state in momentum space by an amount $m\mathbf{v}_s$. We now inquire to what extent such a wave-function corresponds to a *metastable equilibrium* of the liquid, displaying the characteristic feature of "resistance free flow." More explicitly, we ask whether and how the pipe walls could slow down superfluid flow. We shall thus encounter the important concept of a *critical velocity*, above which superfluid flow becomes unstable.

Let us consider a simple experiment, in which a Bose liquid flows at a constant velocity through a thin capillary tube. The corresponding superfluid velocity $\mathbf{v}_s$ must be constant throughout the cross-section of the tube. The superfluid motion is thus a *uniform* translation, in contrast with ordinary viscous flow, for which the fluid velocity varies from the walls to the center of the tube.

It is clear that any interaction of the fluid with the pipe walls cannot change the momentum of the condensed phase, as this would involve a simultaneous transition of the whole liquid, which is highly improbable. Hence, the only way by which the walls can slow down the flow is by creating elementary excitations, thereby absorbing momentum and energy from the uniform motion of the liquid. We are thus led to study the extent to which the walls can create such excitations.

We may safely assume that the walls are infinitely heavy. Therefore, in the frame of reference in which they are fixed, they transfer momentum, but not energy to the system (a point which we have earlier discussed in connection with the rotating bucket experiment). It is clear that the walls cannot create quasi-particles as long as the latter have a *positive* energy in the pipe frame of reference (since at $T = 0$ there is nothing to provide the required energy). Under such conditions, we expect superfluid flow to be stable—at least against quasi-particle creation. If, on the other hand, there exist quasi-particle states whose energy in the pipe frame is negative, multiple excitation processes become possible; they act to damp superfluid motion.

Thus far we only know the excitation spectrum of the Bose liquid in the frame of reference in which the condensed phase is at rest (with zero relative momentum). In that "condensate" frame (which has a velocity $\mathbf{v}_s$), the spectrum is just that found in the preceding section. The spectrum in the pipe frame may be obtained by means of a simple Galilean transformation. Let $\mathbf{p}_i$ be the momentum of a fluid particle, of mass m, as measured in the condensate frame. In the pipe frame, the system Hamiltonian is:

$$
\begin{aligned}
H &= \sum_i \frac{(\mathbf{p}_i + m\mathbf{v}_s)^2}{2m} + \frac{1}{2}\sum_{i\neq j} V(\mathbf{r}_i - \mathbf{r}_j) \\
&= \sum_i \frac{p_i^2}{2m} + \frac{1}{2}\sum_{i\neq j} V(\mathbf{r}_i - \mathbf{r}_j) + \mathbf{p}\cdot\mathbf{v}_s + \frac{Nmv_s^2}{2} \qquad (5.52)
\end{aligned}
$$

where $\mathbf{p} = \sum_i \mathbf{p}_i$ is the net momentum of the fluid particles, as measured in the condensate frame. If, first, we consider the fluid to be in its ground state in the condensate frame (with energy E_o, momentum $\mathbf{p} = 0$) the energy of the system as measured in the pipe frame is

$$E_o + \frac{Nmv_s^2}{2} \qquad (5.53)$$

In the same way, if we assume there is present a single quasi-particle, momentum $\mathbf{p}$, energy ε_p, as measured in the condensate frame, the fixed observer, tied to the pipe, will measure a system energy which is

$$E_o + \varepsilon_p + \mathbf{v}_s \cdot \mathbf{p} + \frac{Nmv_s^2}{2} \tag{5.54}$$

It follows that in the pipe frame the quasi-particle has an energy

$$\varepsilon_p + \mathbf{p} \cdot \mathbf{v}_s \tag{5.55}$$

Thus the walls cannot create quasi-particles as long as

$$\varepsilon_p + \mathbf{p} \cdot \mathbf{v}_s > 0 \text{ for all values of } \mathbf{p} \tag{5.56}$$

If the condition (5.56) is met, we expect superfluid flow to be ***stable.*** Such a criterion was first formulated by Landau in his early work on liquid helium.

Let us suppose that the condition (5.56) fails to be satisfied for some particular values of $\mathbf{p}$ and $\mathbf{v}_s$. In this case, the walls keep creating new excitations of wave vector $\mathbf{p}$. As a result, the particular quasi-particle mode will grow exponentially in time, until that growth is limited by non-linear effects. The fluid flow is ***unstable***, in that there is a steady transfer of energy and momentum from the coherent, directed motion of the condensate to an essentially incoherent group of quasi-particle modes. The instability corresponds to a transformation of the directed kinetic energy into heat, a phenomenon which is typical of viscous damping of fluid flow.

Such an instability, characterized by the sudden onset of viscosity, will occur when the liquid velocity exceeds a critical velocity, v_c, which is given by

$$v_c = \text{lower limit of } \frac{\varepsilon_p}{p} \tag{5.57}$$

For $v_s < v_c$, there is no mechanism by which the fluid flow can transform its kinetic energy into heat: the flow is ***superfluid*** in character, being characterized by a complete absence of any viscosity.

According to the Landau criterion, (5.57), a free Bose gas should not be superfluid. In that case, ε_p is equal to $p^2/2m$: for an arbitrarily low velocity $\mathbf{v}_s$, one can always find a small enough value of $\mathbf{p}$ such that (5.56) is violated. Such a conclusion is at first sight surprising, since we have seen in Chapter 4 that a free Bose gas did ***not*** respond to a macroscopic transverse probe; in that respect, it behaves like a genuine superfluid. Thus, there appears to be a contradiction between the two criteria for superfluid behavior, (4.10) and (3.5). The answer to this paradox is simply that the ground state of the free Bose gas is *superfluid* [as shown by (4.10)], while it is *unstable* against any motion of the fluid, however slow. This instability arises as a consequence of the parabolic nature of the excitation spectrum, and may be traced back to the absence of compressibility in the free Bose gas. The latter thus represents a very "pathological" case, which actually is very sensitive to boundary conditions (it may be shown that when the criterion (5.56) is expressed in terms of the true eigenstates of the liquid in the capillary tube, it is satisfied for low enough velocities $\mathbf{v}_s$). Such difficulties do not arise in the real case of an interacting Bose liquid: however weak the interaction, it will always give rise to a finite compressibility and sound velocity. The slope of the quasi-particle spectrum near the origin is then finite, so that the critical velocity v_c, defined by (5.57), no longer vanishes.

We carried out the previous analysis in a fixed frame of reference, tied to the walls of the system. It is not uninteresting to consider instead the problem from the vantage point of the moving fluid. The walls then appear as massive obstacles which move at velocity $-\mathbf{v}_s$ relative to the fluid. The persistence of superfluid flow then depends on the ability of these moving obstacles to scatter against the liquid.

Instead of a wall, let us consider a massive obstacle of microscopic size, such that its scattering against the fluid can be treated within the Born approximation. Such an obstacle behaves as a test particle which is coupled to the density fluctuations in the liquid. According to the general discussion given in Chapter 2, Vol. I, the probability per unit time that the object transfers momentum $\mathbf{p}$ and energy

$$\frac{(M\mathbf{v}_s + \mathbf{p})^2}{2M} - \frac{Mv_s^2}{2} = \mathbf{p} \cdot \mathbf{v}_s \tag{5.58}$$

to the liquid is proportional to the dynamic form factor

$$S(\mathbf{p}, \mathbf{p} \cdot \mathbf{v}_s) \tag{5.59}$$

[see (I.2.11)]. If spontaneous creation of density fluctuation excitations in the moving fluid is to be forbidden, then the dynamic form factor for the fluid must satisfy the condition:

$$S(\mathbf{p}, \mathbf{p} \cdot \mathbf{v}_s) = 0 \tag{5.60}$$

Equation (5.60) represents an obvious generalization of the Landau criterion, (5.56). It governs not only that part of the density fluctuation excitation spectrum which corresponds to the creation of a single quasi-particle excitation, but also the creation of multi-particle excitations. An advantage of (5.60) is that it provides an exact form of the stability criterion even if the single quasi-particle excitations are damped, since (5.60) governs as well the states into which the excitation decays.

Actually, the Landau criterion (5.56) is not a sufficient condition to observe superfluidity. It clearly represents a necessary condition, which prevents the spontaneous excitation of quasi-particles by the moving walls. However, it does not preclude the existence of other excitations, of a lower energy, which would be excited at lower superfluid velocities. We shall see in Chapter 8 that such excitations do in fact exist. They involve a vortex-like motion of the superfluid, leading to a sort of superfluid *turbulence.* Such a turbulence appears at velocities v_s much smaller than that given by (5.57), and thus controls the stability of superfluid flow.

The Landau criterion, (5.56), is nevertheless very important, as it clearly displays the physical origin of superfluid behavior, namely the scarcity of low lying excited states (itself a consequence of Bose condensation). Furthermore, the critical velocity (5.57) marks the onset of *viscosity* in a superfluid Bose liquid (the vortex motion mentioned above corresponds to a turbulent non-viscous flow). The properties discussed in the present chapter are thus essential to our understanding of superfluidity.

5.6 Condensate Response as Superfluid Motion

In the course of this chapter, we have mentioned several times the intimate relationship between long wave length phonons and macroscopic superfluid motion. Such phonons involve a *longitudinal* superfluid flow, associated with density fluctuations [see (5.18)]. In order to show this connection more clearly, we consider the response to a weak scalar potential (i.e., to a test charge probe). The probe is assumed to be periodic in space and time, with wave vector $\mathbf{q}$ and frequency ω. The perturbing Hamiltonian may thus be written as

$$H_{\text{ext}} = \alpha\left[\rho_{\mathbf{q}}^{+} e^{-i\omega t} + \rho_{\mathbf{q}} e^{i\omega t}\right] \tag{5.61}$$

(corresponding to an applied potential $\alpha\cos(\mathbf{q}\cdot\mathbf{r}-\omega t)$). The perturbation (5.61) is assumed to act on the ground state ψ_o of the Bose liquid.

Using elementary first order perturbation theory, we may write the perturbed wave-function as

$$\psi = \psi_o - \alpha \sum_n \left\{ \frac{(\rho_{\mathbf{q}}^+)_{no}\, e^{-i\omega t}}{\omega_{no} - \omega} + \frac{(\rho_{\mathbf{q}})_{no}\, e^{i\omega t}}{\omega_{no} + \omega} \right\} \psi_n \tag{5.62}$$

In the long wave length limit, the wave-functions $\rho_{\mathbf{q}}^+|0\rangle$ are *eigenstates* of the system, obtained by creating one phonon with wave vector **q**, energy ε_q. The perturbed wave-function may thus be cast in the form

$$\psi = 1 - \alpha \left[\frac{\rho_{\mathbf{q}}^+ e^{-i\omega t}}{\varepsilon_q - \omega} + \frac{\rho_{\mathbf{q}}\, e^{i\omega t}}{\varepsilon_q + \omega} \right] \psi_o \tag{5.63}$$

(from now on, we return to our usual convention of setting $\hbar = 1$). Equation (5.63) clearly describes an admixture of phonons into the ground state ψ_o. In other words, the probe acts to *excite* phonons out of the condensate. Let us replace $\rho_{\mathbf{q}}^+$ by its expression

$$\rho_{\mathbf{q}}^+ = \sum_i e^{+i\mathbf{q}\cdot\mathbf{r}_i}$$

The wave-function (5.63) may then be written as

$$\psi = \left\{ 1 - \frac{2\alpha}{\varepsilon_q^2 - \omega^2} \sum_i \left[\varepsilon_q \cos\left(\mathbf{q}\cdot\mathbf{r}_i - \omega t\right) + i\omega \sin\left(\mathbf{q}\cdot\mathbf{r}_i - \omega t\right) \right] \right\} \psi_o \tag{5.64}$$

In that form, we see that ψ is the expansion up to *first order* in α of the extended London wave-function (5.15), with

$$f = 1 - \frac{2\alpha\varepsilon_q}{\varepsilon_q^2 - \omega^2} \cos(\mathbf{q}\cdot\mathbf{r} - \omega t) \tag{5.65a}$$

$$S = -\frac{2\alpha\varepsilon_q}{\varepsilon_q^2 - \omega^2} \sin(\mathbf{q}\cdot\mathbf{r} - \omega t) \tag{5.65b}$$

The system response to the applied probe may thus be viewed as a small amplitude superfluid motion involving current and density fluctuations determined by S and f respectively. We thus see clearly the dual aspects of long wave length phonons; they represent on the one hand "elementary excitations," on the other "oscillatory superfluid motion." (Such a neat interpretation is valid only in the macroscopic limit.)

According to (5.26), the superfluid velocity $\mathbf{v}_s(\mathbf{r})$ associated with the wave-function (5.64) is equal to

$$\mathbf{v}_s = -\frac{1}{m}\frac{2\alpha\omega}{\varepsilon_q^2 - \omega^2}\mathbf{q}\cos(\mathbf{q}\cdot\mathbf{r} - \omega t) \tag{5.66}$$

We note that $\mathbf{v}_s$ is longitudinal in character (in that $\mathbf{div}\ \mathbf{v}_s \neq 0$): we verify explicitly our earlier statement that such fields correspond to virtual phonon emission. The corresponding density fluctuations may be obtained from (5.30) and equation (5.65a).

$$\delta\rho(\mathbf{r}) = -\frac{N\varepsilon_q}{ms^2}\frac{2\alpha\varepsilon_q}{\varepsilon_q^2 - \omega^2}\cos(\mathbf{q}\cdot\mathbf{r} - \omega t) \tag{5.67}$$

Since the phonon energy ε_q is equal to sq, we may also write

$$\delta\rho(\mathbf{r}) = -\frac{Nq^2}{m}\frac{2\alpha}{\varepsilon_q^2 - \omega^2}\cos(\mathbf{q}\cdot\mathbf{r} - \omega t) \tag{5.68}$$

Such a result could in fact have been obtained directly by noting that

$$\delta\rho(\mathbf{r}) = 2\alpha\chi(\mathbf{q},\omega)\cos(\mathbf{q}\cdot\mathbf{r} - \omega t) \tag{5.69}$$

where $\chi(\mathbf{q},\omega)$, the density-density response function, is given by (I.2.69) and (2.28).

It follows from (5.66) and (5.68) that

$$N\mathrm{div}\ \mathbf{v}_s + \frac{\partial\delta\rho}{\partial t} = 0 \tag{5.70}$$

which, of course, is the usual *continuity equation* (expanded up to first order in α). Moreover, we obtain from (5.66) and (5.67) the following relation

$$\frac{\partial\mathbf{v}_s}{\partial t} + \frac{s^2}{N}\mathbf{grad}\ (\delta\rho) = \frac{2\alpha}{m}\mathbf{q}\sin(\mathbf{q}\cdot\mathbf{r} - \omega t) = -\frac{1}{m}\mathbf{grad}\ \varphi \tag{5.71}$$

where

$$\varphi = 2\alpha\cos(\mathbf{q}\cdot\mathbf{r} - \omega t) \tag{5.72}$$

is the potential exerted by the external probe on the system particles. (5.71) is the *dynamical equation* governing the time dependence of superfluid motion. It can be made even more explicit if we notice that a density fluctuation $\delta\rho$ gives rise to a shift in the chemical potential

$$\delta\mu = \frac{ms^2}{N}\delta\rho \tag{5.73}$$

[see for instance (I.1.49)]. (5.71) may then be written as

$$m\frac{\partial \mathbf{v}_s}{\partial t} = -\mathbf{grad}\ (\mu + \varphi) \tag{5.74}$$

(5.74) is the *linearized equation of motion for superfluid flow.* It clearly shows that superfluid acceleration arises as a consequence of both an "internal" force, $-\mathbf{grad}\,\mu$, and an "external" force, $-\mathbf{grad}\ \varphi$.

The results (5.70) and (5.74) provide a macroscopic description of the dynamical properties of a superfluid Bose liquid at zero temperature. When extended to finite temperatures, they permit one to discuss non-equilibrium phenomena in the framework of the two-fluid model (see Chapter 7). We shall now give a direct proof of these important equations.

5.7 Dynamical Properties of Condensate Motion

Let us consider an isolated Bose liquid, which is not acted on by any external force (i.e., $\varphi = 0$). The liquid is assumed to be in metastable equilibrium at zero temperature. It is characterized by a slowly varying time-dependent condensate wave-function $\phi(\mathbf{r}, t)$, which we write in the form (5.16). At every point $\mathbf{r}$, the liquid possesses a velocity $\mathbf{v}_s(\mathbf{r})$ and a density fluctuation $\delta\rho(\mathbf{r})$, which are given respectively by (5.26) and (5.30). We wish to find the equations satisfied by these two quantities.

The continuity equation (5.70) follows at once from the general conservation law, (I.2.39), together with relation (5.25). On the other hand, the dynamical equation (5.74) is far less obvious, and requires a detailed proof. For that purpose, we consider the quantity

$$I = \int d^3\mathbf{r}_1 \ldots d^3\mathbf{r}_n \left\{ \psi^*(\mathbf{r}_1 \ldots \mathbf{r}_n, t)\frac{\partial}{\partial t}\psi(\mathbf{r}_1 \ldots \mathbf{r}_n, t) - \psi(\mathbf{r}_1 \ldots \mathbf{r}_n, t)\frac{\partial}{\partial t}\psi^*(\mathbf{r}_1 \ldots \mathbf{r}_n, t)\right\} \tag{5.75}$$

We first replace ψ by its expression (5.19). On noting that

$$\frac{\partial \psi_o}{\partial t} = -iH\psi_o = -iE_o\psi_o \tag{5.76}$$

(where E_o is the ground state energy), we easily reduce (5.75) to the form

$$I = 2i\int d^3\mathbf{r}_1 \ldots d^3\mathbf{r}_n\,|\psi|^2 \left\{ \sum_i \frac{\partial S(\mathbf{r}_i, t)}{\partial t} - E_o \right\} \tag{5.77}$$

Equation (5.77) may be further simplified by referring to the definition (5.20) of the density $\rho(\mathbf{r})$; we thus find

$$I = 2i \int d^3\mathbf{r}\rho(\mathbf{r})\frac{\partial S(\mathbf{r},t)}{\partial t} - 2i \int d^3\mathbf{r}_1 \dots d^3\mathbf{r}_n \psi^* E_o \psi \tag{5.78}$$

On the other hand, ψ satisfies the Schrödinger equation

$$i\frac{\partial \psi}{\partial t} = H\psi \tag{5.79}$$

Inserting (5.79) into (5.75), we obtain

$$I = 2i \int d^3\mathbf{r}_1 \dots d^3\mathbf{r}_n \psi^* H \psi \tag{5.80}$$

Comparison of (5.78) with (5.80) shows that

$$\int d^3\mathbf{r}\rho(\mathbf{r})\frac{\partial S(\mathbf{r},t)}{\partial t} = -\int d^3\mathbf{r}_1 \dots d^3\mathbf{r}_n \psi^* \left(H - E_o\right)\psi \tag{5.81}$$

In practice, the wave-function ψ is normalized up to second order in the density fluctuations. To that same accuracy, we can thus write

$$\int d^3\mathbf{r}\rho(\mathbf{r})\frac{\partial S(\mathbf{r},t)}{\partial t} = -\left(E - E_o\right) \tag{5.82}$$

where E is the energy corresponding to the wave-function ψ. (5.82) is an "integrated" version of the dynamical equation for S.

Let us make use of the expression (1.133) for the energy difference $(E - E_o)$. (5.82) thus becomes

$$\int d^3\mathbf{r} \left\{\rho(\mathbf{r})\left[\frac{\partial S(\mathbf{r},t)}{\partial t} + \frac{1}{2}m\left[\mathbf{v}_s(\mathbf{r},t)\right]^2\right] + \frac{ms^2}{2N}\delta\rho(\mathbf{r},t)^2\right\} = 0 \tag{5.83}$$

We now consider a small fictitious local increase of the actual density $\rho(\mathbf{r})$; $\delta\rho(\mathbf{r})$ clearly increases by the same amount. Since the left hand side of (5.83) must remain unchanged by such an increase, it follows that

$$\frac{\partial S(\mathbf{r},t)}{\partial t} + \frac{1}{2}m\left[\mathbf{v}_s(\mathbf{r},t)\right]^2 + \frac{ms^2}{N}\delta\rho(\mathbf{r},t) = 0 \tag{5.84}$$

According to (5.73), the last term on the left hand side of (5.84) is the shift in the chemical potential, $\delta\mu(\mathbf{r},t)$, associated with the superfluid motion.

(5.84) determines the time dependence of the phase S. On taking its gradient in ordinary space, and making use of (5.26), we find

$$m\left[\frac{\partial \mathbf{v}_s}{\partial t}+\frac{1}{2}\mathbf{grad}\ (\mathbf{v}_s)^2\right]=-\mathbf{grad}\ \mu \tag{5.85}$$

Since curl $\mathbf{v}_s = 0$, we may write (5.85) as

$$m\frac{\partial \mathbf{v}_s}{\partial t}+(\mathbf{v}_s\cdot\mathbf{grad})\,\mathbf{v}_s=m\frac{d\mathbf{v}_s}{dt}=-\mathbf{grad}\ \mu \tag{5.86}$$

where d/dt denotes the usual total derivative. (5.86) is the basic equation of motion for superfluid flow. The previous result (5.74) taken for $\varphi = 0$, appears as a linearized version of (5.86), valid for moderate values of $\mathbf{v}_s$. Again, we note that μ acts as the potential felt by the condensate.

Equations (5.70) and (5.86) may be used to set up a macroscopic theory of sound wave propagation at $T = 0$. One finds, of course, phonons which propagate with velocity s.

CHAPTER 6

BASIS FOR THE TWO-FLUID MODEL [1]

6.1 On the Calculation of ρ_n

We now consider the behavior of the superfluid Bose liquid at finite temperatures. We have remarked in Chapter 1 that the two-fluid theory of Tisza and Landau provides an excellent macroscopic description of such superfluid behavior. It is natural to inquire whether the Bose liquid theory we have developed in Chapters 2–5 permits a precise definition of the basic quantities which appear in the two-fluid model. The present chapter is devoted primarily to answering that question; in so doing we shall be led to discuss the extent to which the usual two-fluid equations may be expected to be applicable. We shall confine our attention to equilibrium situations, postponing to the following chapter consideration of a non-equilibrium case.

We shall follow Landau, and assume that the differences between the Bose liquid behavior at finite temperatures and that at $T = 0$ arise solely as a consequence of the thermal excitation of quasi-particles. For $T < T_c$ there will continue to be macroscopic occupation of a single quantum state. That state, the condensate, is identified as the "superfluid" component of the two-fluid model. As at $T = 0$, it displays the characteristic superfluid properties of *irrotational, reversible, resistance-free flow.* The thermally-excited quasi-particles are identified with the normal fluid. In

[1]In order to conform with the standard notation in the literature, throughout the next two chapters we shall use ρ and J to represent *mass* density and *mass* current density, rather than number density and current density, as has hitherto been the case. It is hoped this change in notation will cause no great amount of confusion to the reader.

general their motion will be irreversible and may be described, in suitable limit, with the aid of a coefficient of viscosity. They scatter against the walls of the container and, for a sufficiently small container, may be regarded as being in equilibrium with it. Moreover, their response to external probes will be seen to be "ordinary." Thus the thermal quasi-particles respond symmetrically to low-frequency, long wave-length, transverse and longitudinal probes.

The basic variables of the two-fluid theory are ρ_s, $\mathbf{v}_s$, ρ_n, and $\mathbf{v}_n$, the mass density and velocity of the superfluid and normal components, respectively, of the liquid. ρ_n and ρ_s are not independent, their sum being the total mass density of the fluid,

$$\rho = \rho_n + \rho_s \tag{1.1}$$

while the mass current in the liquid is taken to be

$$\mathbf{J} = \rho_n \mathbf{v}_n + \rho_s \mathbf{v}_s \tag{1.2}$$

These equations, together with (1.3), which specifies that the total entropy is carried by the normal component, serve to characterize an equilibrium situation, in which $\mathbf{v}_n$ and $\mathbf{v}_s$ are constant, while ρ_n and ρ_s depend only on temperature. We have a definite prescription for the calculation of the condensate velocity, $\mathbf{v}_s$. According to (5.26) it is proportional to the gradient of the phase of the condensate wave function, $\phi(\mathbf{r})$. We shall see that $\mathbf{v}_n$ is, by contrast, only defined through a statistical "averaging" process; it corresponds to an "average" velocity of the thermally-excited quasi-particles. Both $\mathbf{v}_s$ and $\mathbf{v}_n$ are determined by the external boundary conditions appropriate to the physical situation under discussion. In order to have a specific microscopic basis for the above two-fluid equations we need seek only an unambiguous prescription for one of the two remaining variables, say ρ_n.

Consideration of the rotating bucket experiment provides such a prescription, both experimentally and theoretically. Since condensate motion is irrotational, the superfluid component will not respond to rotation of its container. On the other hand, the quasi-particles, assumed to be in equilibrium with the bucket walls, do respond. The moment of inertia of the liquid in the bucket is then determined entirely by the normal fluid, and is directly proportional to ρ_n. As emphasized by Landau (1941), an Andronikashvili-type rotating disc experiment thus provides a direct measurement of ρ_n.

A theoretical definition of ρ_n may be obtained by analyzing the experiment from a response-function point of view, and then comparing the results of that analysis with the predictions of the two-fluid model. According to the two-fluid model, the normal fluid will follow the rotation of the bucket at frequency ω. The local velocity of the normal fluid at a distance r from the axis of rotation is thus

$$\mathbf{v}_n(\mathbf{r}) = \boldsymbol{\omega} \times \mathbf{r} \tag{6.1}$$

The superfluid component does not follow so that $v_s = 0$. The mass current is therefore given by

$$\mathbf{J}(\mathbf{r}) = \rho_n \boldsymbol{\omega} \times \mathbf{r} \tag{6.2}$$

where ρ_n is a constant if $\mathbf{v}_n(\mathbf{r})$ varies sufficiently slowly in space.

On the other hand, slow rotation of the container serves as a transverse probe of the liquid contained within; we have seen in Chapter 4 that the entire system responds as if it were subjected to a vector potential:

$$\mathbf{A}_\perp(\mathbf{r}) = \boldsymbol{\omega} \times \mathbf{r} \tag{6.3}$$

Since the condensate does not respond, the resulting mass current is due entirely to the thermal quasi-particles. It is determined by the finite temperature transverse current-current response function. If $\mathbf{A}_\perp(\mathbf{r})$ varies sufficiently slowly in space, this response function is a constant, so that we can write [cf. (4.9)]

$$\mathbf{J}(\mathbf{r}) = -m^2 \chi_\perp \boldsymbol{\omega} \times \mathbf{r} \tag{6.4}$$

Comparison of (6.2) and (6.4) offers the desired specification of ρ_n in terms of the appropriate transverse current response function.

The resulting expression for ρ_n has the great advantage both of being exact and of corresponding directly to what is measured experimentally. However, it does not obviously lend itself to ready calculation. Landau obtained another approximate expression for ρ_n by considering the two-fluid description of a different experiment, that of flow of the liquid through a narrow pipe. Under circumstances such that the thermal quasi-particles are scattered frequently by the walls of the pipe, they will come to equilibrium with them. Collisions with the walls thus represent the physical mechanism by which the normal fluid (the quasi-particles) is singled out, the superfluid component being unaffected by the presence of the pipe.

Suppose the velocity of the condensate is $\mathbf{v}$. According to the two-fluid model, the velocity of the normal fluid vanishes, since it is "attached" to the walls of the pipe by viscous effects. Now let us go to a frame of reference which moves with the condensate. In this frame, the superfluid velocity, $\mathbf{v}_s$, vanishes, while the normal fluid will possess a velocity:

$$\mathbf{v}_n = -\mathbf{v} \tag{6.5}$$

The corresponding mass current is

$$\mathbf{J} = -\rho_n \mathbf{v}_n = \rho_n \mathbf{v} \tag{6.6}$$

The mass current $\mathbf{J}'$, in the pipe frame, is:

$$\mathbf{J}' = -\rho_n \mathbf{v}_n + \rho \mathbf{v} = \rho_s \mathbf{v} \tag{6.7}$$

At $T = 0$, it would be $\rho\mathbf{v}$; we see that thermal quasi-particles in equilibrium act effectively to reduce the ***strength*** of the superfluid flow, while not affecting its absence of viscosity.

If the thermal quasi-particles are regarded as a non-interacting excitation gas, one may readily use considerations based on Galilean invariance to calculate the current carried by the quasi-particles, $\mathbf{J}'$. The resulting expression for ρ_n, first obtained by Landau, is, as we shall see, not quite exact, since in fact the quasi-particles form an interacting excitation liquid, rather than a non-interacting excitation gas. It is, nonetheless, extremely useful.

We now consider in more detail the above specifications of ρ_n, and the extent to which they are related.

6.2 A Response-Function Definition of ρ_n

We have seen that rotation at angular velocity ω may be formally described as a perturbing term in the system Hamiltonian:

$$-\boldsymbol{\omega} \cdot \mathbf{L} = -m \sum_{\mathbf{q}} \mathbf{j_q}^{+} \cdot \mathbf{A_q} \tag{6.8}$$

where $\mathbf{j_q}^{+}$ is the ***particle*** current density fluctuation and $\mathbf{A_q}$ is the Fourier-transform of the transverse "vector potential," (4.7). Application of the standard response-function theory shows that this perturbation gives rise to a ***mass*** current in the condensate (fixed) frame whose Fourier-component is

$$\langle \mathbf{J_q} \rangle = -m^2 \chi_\perp(\mathbf{q}, o) \mathbf{A_q} \tag{6.9}$$

$\chi_\perp(\mathbf{q}, o)$ is the finite temperature current-current response function, specified by

$$\chi_\perp(\mathbf{q}, o) = 2Z^{-1} \sum_{nm} \frac{e^{-\beta E_m} \left| \left(\mathbf{j_q} \cdot \boldsymbol{\eta}_{\mathbf{q}\perp} \right)_{nm} \right|^2}{\omega_{nm}} \tag{6.10}$$

To the extent that $\mathbf{A}_\perp(\mathbf{r})$ varies sufficiently slowly in space, $\chi_\perp(\mathbf{q}, o)$ is independent of $\mathbf{q}$ for all the values, $\mathbf{A_q}$, which appear. Under these circumstances the Fourier transform of (6.9) is simply (6.4); on comparing this with (6.2), we may therefore write:

$$\begin{aligned} \rho_n &= -m^2 \lim_{\mathbf{q}\to 0} \chi_\perp(\mathbf{q}, o) \\ &= -\frac{2m^2}{Z} \lim_{\mathbf{q}\to 0} \sum_{mn} \frac{e^{-\beta E_m} \left| \left(\mathbf{j_q}^+ \cdot \boldsymbol{\eta}_{\mathbf{q}\perp} \right)_{nm} \right|^2}{\omega_{nm}} \end{aligned} \tag{6.11}$$

Equation (6.11) provides the desired definition of ρ_n in terms of the exact matrix elements and excitation frequencies appropriate to the Bose liquid. It is an enormously appealing expression, because it has built into it the following feature: *if superfluidity is defined as the presence of long-range correlations in the current-current response function, then the transition from the superfluid to the normal state automatically takes place when $\rho_n = \rho$.* This important result follows at once from the definition of ρ_n and $\chi_\perp(\mathbf{q}, o)$. We have seen that long-range correlations cease to occur at temperature T such that

$$\lim_{\mathbf{q}\to 0} \chi_\perp(\mathbf{q}, o) = \lim_{\mathbf{q}\to 0} \chi_\parallel(\mathbf{q}, o) \tag{6.12}$$

On the other hand, the f-sum rule tells us that

$$\chi_\parallel(\mathbf{q}, o) = -\frac{N}{m} \tag{6.13}$$

at all temperatures and wave lengths. It follows at once that the equality (6.12) is equivalent to the equality,

$$\rho_n = \rho$$

Without any detailed calculation we can see that according to (6.11) the only non-vanishing contributions to ρ_n come from transitions between states which involve thermally-excited quasi-particles. Exactly as at zero temperature in the long wave-length limit, the only non-negligible matrix elements of the current density fluctuation *when acting on the condensate* are those for the excitation or de-excitation of a *single* quasi-particle from the condensate. For reasons of symmetry, the corresponding transverse matrix element vanishes. Hence the condensate does not respond to the rotation of the container.

On the other hand, the *thermally excited quasi-particles* will respond to such a rotation. One may have scattering of an already excited quasi-particle of momentum $\mathbf{p}$ to a state of momentum $\mathbf{p}+\mathbf{q}$. There is no particular symmetry in such a scattering act, so that the corresponding "transverse" matrix element need not vanish. Moreover, the excitation frequency going with such a scattering, $\varepsilon_{\mathbf{p+q}} - \varepsilon_{p}$, may be arbitrarily small in the long wave-length limit, so that such a virtual scattering-process contributes to $\chi_{\perp}(\mathbf{q}, o)$.

We may inquire just how small $\mathbf{q}$ must be in order that the above definition of ρ_n in terms of the transverse response function will apply. Clearly the definition is satisfactory provided there exists a *local* relation between the induced current, $\mathbf{J}(\mathbf{r})$, and the perturbing "potential," $\mathbf{v}_n(\mathbf{r})$, as specified by (6.2). We have seen in Chapter 4 that such a local relation may be expected provided the external probe varies slowly over a coherence length. For liquid He II, the coherence length is of the order of the inter-particle spacing. It follows that as long as

$$qr_o \ll 1 \tag{6.14}$$

the relation, (6.11), will be valid.

6.3 Superfluid Flow

We now consider the Landau definition of ρ_n, which is based on analysis of the flow, at velocity $\mathbf{v}$, of the Bose liquid through a narrow pipe. According to (6.6), ρ_n may be readily calculated if one knows how to calculate the mass current density, J, in the "condensate" frame of reference (that moving with the condensate), since in this frame, the entire current is carried by the thermal quasi-particles. Let us suppose, for the moment, that quasi-particles form a non-interacting gas. Their energy, ε_p, in the condensate frame is the same as at equilibrium, since in that frame the condensate is at rest. The corresponding energy, $\varepsilon'_{\mathbf{p}}$, as viewed in the "pipe" frame is equal to

$$\varepsilon'_{\mathbf{p}} = \varepsilon_p + \mathbf{p}\cdot\mathbf{v} \tag{6.15}$$

Since the quasi-particles are in equilibrium with the pipe walls (which act as a thermostat fixing the temperature T), the Bose-Einstein equilibrium distribution $n_{\mathbf{p}}(T, \mathbf{v})$ at temperature T will involve the energy $\varepsilon'_{\mathbf{p}}$, not ε_p. It is:

$$n_{\mathbf{p}}(T, \mathbf{v}) = \frac{1}{e^{(\varepsilon_p + \mathbf{p}\cdot\mathbf{v})} - 1} = n^o_{\mathbf{p}}(\varepsilon_p + \mathbf{p}\cdot\mathbf{v}) \tag{6.16}$$

As a consequence of superfluid flow, the thermal quasi-particle distribution is asymmetric in momentum space. Such asymmetry gives rise to a net mass current in the condensate frame, which may be written as

$$\mathbf{J} = \sum_{\mathbf{p}} n_{\mathbf{p}}(T, \mathbf{v})\mathbf{p} = \sum_{\mathbf{p}} \mathbf{p} n^o_{\mathbf{p}}(\varepsilon_p + \mathbf{p}\cdot\mathbf{v}) \tag{6.17}$$

We assume $\mathbf{v}$ to be small. $\mathbf{J}$ is then given accurately by the lowest order non-vanishing term,

$$\mathbf{J} = \sum_{\mathbf{p}} \mathbf{p}\mathbf{p}\cdot\mathbf{v}\frac{\partial n^o_{\mathbf{p}}}{\partial \varepsilon_p} = \frac{1}{3}\sum_{\mathbf{p}} p^2 \frac{\partial n^o_{\mathbf{p}}}{\partial \varepsilon_p}\mathbf{v} \tag{6.18}$$

on making use of the symmetry about $\mathbf{v}$. If we now compare (6.18) with (6.6), we see that

$$\rho_n = -\frac{1}{3}\sum_{\mathbf{p}} p^2 \frac{\partial n_p}{\partial \varepsilon_p} = +\frac{1}{3}\sum_{\mathbf{p}} p^2 \frac{e^{\beta\varepsilon_p}}{(e^{\beta\varepsilon_p} - 1)^2} \tag{6.19}$$

Equation (6.19) is the Landau definition of ρ_n in terms of the distribution function for the thermally-excited quasi-particles. We have mentioned that it is not an exact expression, because the thermal quasi-particles form an interacting excitation liquid, rather than a non-interacting excitation gas. The point is that, as mentioned in Chapter 2, the energy of a quasi-particle, ε_p, in the condensate frame of reference is a *functional* of the distribution function, n_p, in that frame. [Thus, when we employed the usual equilibrium distribution function, $n_p(T)$, ε_p turned out to depend on the temperature.] In the present case, the quasi-particle distribution function depends on $\mathbf{v}$, since the quasi-particles are in thermal equilibrium with the walls. Such a dependence on $\mathbf{v}$ will in turn act back on ε_p, so that ε_p becomes a function of both T and $\mathbf{v}$. In short, moving the quasi-particles with respect to the condensate changes the distribution function n_p, which in turn changes the energy ε_p, thence the equilibrium distribution, and finally the current $\mathbf{J}'$.

If one expands the quasi-particle current, $\mathbf{J}$, to first order in $\mathbf{v}$, one finds an additional term

$$\sum_{\mathbf{p}\mathbf{p}'} \mathbf{p}\frac{\partial n_p}{\partial \varepsilon_p}\frac{\delta \varepsilon_p}{\delta n_{p'}}\frac{\partial n_{p'}}{\partial \mathbf{v}}\mathbf{v} \tag{6.20}$$

The corresponding correction of ρ_n is not large because it is proportional to the number of thermal quasi-particles present at a given temperature T. It is thus of order N'/N, where

$$N' = \sum_{\mathbf{p}} n_p(\varepsilon_p, T)$$

As we have remarked, for He II N'/N is negligible until one arrives at temperatures quite close to T_c. We consider in a subsequent section the extent to which one finds agreement between experiment and the simple Landau expression, (6.19).

We may now ask whether the Landau value of ρ_n, based on superfluid flow through a pipe, agrees with the definition (6.6), based on a rotating bucket experiment. The equivalence of these two definitions has been established by Balian and de Dominicis (1965), who used field theoretic methods to evaluate the response function in terms of the quasi-particle distribution function; they obtained the Landau result (6.19), supplemented by the above correction.

It is illuminating to carry out an elementary quasi-particle calculation of ρ_n beginning with (6.11). To do this, we note that the current density operator, $\mathbf{j_q}^+$, will act to scatter a quasi-particle of momentum $\mathbf{p}$, energy ε_p, to a new quasi-particle state, of momentum $\mathbf{p}+\mathbf{q}$, energy $\varepsilon_{\mathbf{p}+\mathbf{q}}$. We may write the matrix element between these quasi-particle states as follows:

$$\langle \mathbf{p}+\mathbf{q} | \mathbf{j_q}^+ \cdot \boldsymbol{\eta}_\mathbf{q} | \mathbf{p} \rangle = \xi_{\mathbf{pq}} \frac{\mathbf{p} \cdot \boldsymbol{\eta}_\mathbf{q}}{m} \tag{6.21}$$

where $\xi_{\mathbf{pq}}$ is the coherence factor appropriate to the scattering act in question. It follows, upon application of the statistical considerations of Sec. 2.6, Vol. I, that

$$\begin{aligned}
\rho_n &= -\lim_{\mathbf{q}\to 0} 2 \sum_{p} \frac{n_\mathbf{p}(1+n_{\mathbf{p}+\mathbf{q}})}{\varepsilon_{\mathbf{p}+\mathbf{q}} - \varepsilon_\mathbf{p}} (\mathbf{p}\cdot\boldsymbol{\eta}_\mathbf{q})^2 \xi_{\mathbf{pq}}^2 \\
&= -\lim_{\mathbf{q}\to 0} \sum_{\mathbf{p}} \frac{(n_\mathbf{p} - n_{\mathbf{p}+\mathbf{q}})(\mathbf{p}\cdot\boldsymbol{\eta}_\mathbf{q})^2}{\varepsilon_{\mathbf{p}+\mathbf{q}} - \varepsilon_\mathbf{p}} \xi_{\mathbf{pq}}^2
\end{aligned} \tag{6.22}$$

The second form for ρ_n has been obtained with the aid of the transformation

$$\begin{aligned}
\mathbf{p} &\to -(\mathbf{p}+\mathbf{q}) \\
\mathbf{p}+\mathbf{q} &\to -\mathbf{p}
\end{aligned} \tag{6.23}$$

applied to half the terms in the first form for ρ_n. If we now pass to the limit, $\mathbf{q} \to 0$, in (6.22), we have

$$\rho_n = \lim_{\mathbf{q}\to 0} \sum_{\mathbf{p}} \frac{\mathbf{q}\cdot\nabla_{\mathbf{p}} n_{\mathbf{p}}}{\mathbf{q}\cdot\nabla_{\mathbf{p}}\varepsilon_{\mathbf{p}}} (\mathbf{p}\cdot\eta_{\mathbf{q}})^2\,\xi_{\mathbf{pq}}^2$$

$$= \lim_{\mathbf{q}\to 0} \sum_{\mathbf{p}} \frac{\partial n_{\mathbf{p}}}{\partial \varepsilon_{\mathbf{p}}} (\mathbf{p}\cdot\eta_{\mathbf{q}})^2\,\xi_{\mathbf{pq}}^2 \tag{6.24}$$

On comparing (6.24) with (6.19), we see that the two definitions of ρ_n are identical provided

$$\xi_{\mathbf{pq}} = 1 \tag{6.25}$$

We may thus argue, *a posteriori*, that agreement between the two expressions furnishes a strong argument for believing that all the various renormalization factors which appear in passing from "bare" particles to quasi-particles, simply cancel out.

To conclude this part of our discussion of superfluid flow, we remark that we can pass to the case of the more general two-fluid equation for the current, (1.2), by means of a simple Galilean transformation. Suppose we measure the system current in a frame of reference moving with a velocity $-\mathbf{v}_n$ with respect to the fixed walls of the pipe. In that frame the mass current will be

$$\mathbf{J} = \rho\mathbf{v}_n + \rho_s\mathbf{v} \tag{6.26}$$

since the current measured in the pipe frame of reference is $\rho_s\mathbf{v}$. If we now make use of (1.1), we can write:

$$\mathbf{J} = \rho_n\mathbf{v}_n + \rho_s\mathbf{v}_s \tag{6.27}$$

where

$$\mathbf{v}_s = \mathbf{v} + \mathbf{v}_n \tag{6.28}$$

is the velocity of the condensate measured in our frame of reference moving at velocity $-\mathbf{v}_n$ with respect to the pipe walls. This discussion provides a natural basis for considerations in which it is appropriate to regard the normal fluid, the thermal quasi-particles, as possessing an average drift velocity $\mathbf{v}_n$, while the condensate moves at a relative velocity $\mathbf{v}_s - \mathbf{v}_n$ with respect to the thermal quasi-particles.

6.4 Response to a Longitudinal Probe

Unlike a transverse probe, which acts only on the normal component, a longitudinal probe of system behavior will act on both the condensate and the thermal quasi-particles. It is natural to inquire to what extent one can separate the condensate and thermal quasi-particle contributions to longitudinal response. In this section we shall consider the somewhat formal problem of the system response to a static vector potential coupled to the longitudinal current fluctuations. We consider in Chapter 7 dynamic longitudinal probes of system behavior.

Just as at $T = 0$, there can be net current induced by a longitudinal static vector potential $\mathbf{A_q}$. According to the f-sum rule, the longitudinal current-current correlation function is

$$\chi_{\parallel}(\mathbf{q}, 0) = -\frac{N}{m} \tag{6.29}$$

The induced "paramagnetic" current,

$$\mathbf{J_q} = -m^2 \chi_{\parallel}(\mathbf{q}, o)\mathbf{A_q} = \rho \mathbf{A_q} \; , \tag{6.30}$$

compensates exactly the gauge current, (4.16). At $T = 0$, the correlation function, $\chi_{\parallel}(\mathbf{q}, o)$, is determined entirely by the condensate; it achieves the value (6.29) as a result of virtual excitation of single quasi-particles from the condensate. At finite temperatures this is no longer the case; there will be a contribution to $\chi_{\parallel}(\mathbf{q}, o)$ from the thermal quasi-particles, which is, in fact, $\chi_{\perp}(\mathbf{q}, o)$. To see this, we note that the thermal quasi-particles cannot, in the long wave-length limit, distinguish between a longitudinal and a static transverse probe. In other words, there are no long range correlations present in the "normal" part of the current-current response function.

We therefore write:

$$\lim_{\mathbf{q} \to 0} \chi_{\parallel}(\mathbf{q}, o) = \chi_{\perp}(\mathbf{q}, o) + \chi_s(\mathbf{q}, o) \tag{6.31}$$

where $\chi_s(\mathbf{q}, o)$ specifies the finite-temperature condensate response. On making use of (6.11) and (6.29), we may write

$$\lim_{\mathbf{q} \to 0} \chi_s(\mathbf{q}, o) = -\frac{\rho_s}{m^2} \tag{6.32}$$

The condensate response is thus of strength ρ_s. What we have achieved in the above discussion is a separation of the longitudinal current-current response function into two parts, one proportional to ρ_n, arising from the thermally-excited quasi-particles, the other, proportional to ρ_s, arising from virtual phonon transitions in and out of the condensate. The thermal quasi-particles behave symmetrically in their response to static longitudinal and transverse probes; the condensate responds only to a longitudinal probe.

6.5 Experimental Measurements of ρ_n in He II

There are two well-known methods of measuring ρ_n in He II. One is by the Andronikashvili torsion pendulum experiment; the other is through a measurement of the velocity of second sound. What is of particular interest is the extent to which one finds agreement between the approximate Landau expression for ρ_n, (6.19) and the experimental measurements. The quantity,

$$\delta\rho_n = \rho_n^{\text{Landau}} - \rho_n^{\text{Expt.}} , \tag{6.33}$$

offers a direct measure of the importance of the quasi-particle liquid corrections considered by Balian and de Dominicis. In making the comparison it is important to take account of the fact that one should use, in (6.19), the appropriate quasi-particle energies for the temperature in question; in other words, one should take account of the temperature dependence of ε_p, which we have discussed in Chapter 3. This correction becomes most important near T_c, in a region where the roton energies vary rapidly with temperature. (Throughout the temperature range of interest, rotons near the minimum of the excitation energy curve make the dominant contribution to ρ_n.)

Thus far the most detailed comparison between theory and experiment which has been carried out is due to Bendt, Cowan, and Yarnell (1959). In Fig. 6.1 we reproduce their comparison of the Landau values of ρ_n (including a temperature-dependent ε_p) with experimental measurements. Bendt et al. report that the agreement between their calculations and the values of ρ_n/ρ obtained from second sound measurements is within ±8% over the temperature range 0.7 to 2.0°K, and is within ±5% between 1.1°K and 1.9°K. (The torsion pendulum measurements of ρ_n/ρ are systematically somewhat lower than the second sound measurements.) This measure of agreement may be regarded as excellent, and offers a reliable guide to the extent of departures from the Landau theory at temperatures below 1.8°K. At higher temperatures, the smallness of the $\delta\rho_n$ corrections may perhaps be illusory, because the roton energies used to calculate ρ_n

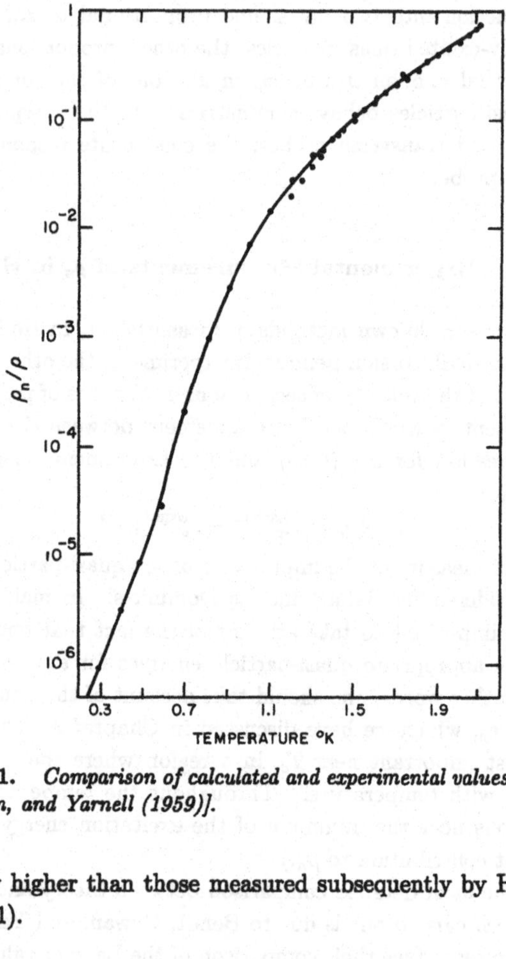

FIGURE 6.1. *Comparison of calculated and experimental values of ρ_n/ρ [from Bendt, Cowan, and Yarnell (1959)].*

lie definitely higher than those measured subsequently by Henshaw and Woods (1961).

CHAPTER 7

FIRST, SECOND, AND QUASI-PARTICLE SOUND

7.1 Collisionless vs. Hydrodynamic Regimes

We now consider the nature of the excited states of a Bose liquid at finite temperatures. In Chapters 2 and 3, we have discussed the quasi-particle excitations at finite temperatures. We here wish to concentrate on the nature of the excited states in the *long wave-length limit.* As in the previous chapters, we shall restrict our attention to states for which the condensate is very nearly uniform.

We recall that at $T = 0$, there exists a unique class of excited states in the long wave-length limit: one finds phonons at velocity s, the macroscopic sound velocity. Such wave propagation might be appropriately called *quasi-particle sound*, since in the long wave-length limit the only density fluctuations of importance are those produced by exciting a single quasi-particle from the condensate. Quasi-particle sound has essentially the same physical origin as zero sound in a Fermi liquid; the restoring force on a given particle comes from the averaged field of all the other particles. (We shall see how this comes about in a microscopic theory in Chapter 9).

At finite temperatures matters are not quite so simple. Quasi-particle sound is still a possible mode of excitation of the Bose liquid; however whether it is found depends on the temperature and the wave length under investigation. As one might expect, the criterion for observing quasi-particle sound is

$$\omega_{\mathbf{q}}\tau_{\mathbf{q}} \gg 1$$

where $\tau_{\mathbf{q}}$ is the life time of the quasi-particle excitation (of energy $\omega_{\mathbf{q}}$) in question. Where this criterion is satisfied, one is effectively in a *collisionless* regime in which the nature of the excitations is little changed from that at $T = 0$. The quasi-particle sound velocity will be temperature dependent, since the energy of a given quasi-particle depends on the thermal excitation of other quasi-particles.

Under most circumstances, the above criterion will not be satisfied in the extreme long wave-length limit: as $q \to 0$, $\varepsilon_q \to 0$, while τ_q will usually remain finite. Hence in the immediate vicinity of $\mathbf{q} = 0$ one is in a *hydrodynamic* regime, in which the restoring force responsible for wave propagation consists in the frequent collisions between thermal excitations which act to bring about local thermodynamic equilibrium . As was the case for the Fermi liquid, one is in the hydrodynamic limit when

$$\omega\tau_r \ll 1$$

where τ_r is the relaxation time required for achieving local thermodynamic equilibrium. It is clear that to the extent that quasi-particle collisions provide the mechanism for maintaining such equilibrium, τ_r will be the same order of magnitude as τ.

We shall be interested in the density fluctuation spectrum in these two limits. We have seen in Sec. 2.6, Vol. I, that the spectrum may be specified by $\chi''(\mathbf{q}, \omega)$, the imaginary part of the density-density response function. In the collisionless regime, we may expect two distinct contributions to $\chi''(\mathbf{q}, \omega)$. One is a quasi-particle sound peak, which arises from excitation and de-excitation of *single* quasi-particles from the condensate. (We consider only wave lengths sufficiently long that multi-particle excitations may be neglected.) The other contribution arises from the scattering of the already thermally-excited quasi-particles. It consists in a continuous spectrum, extending from the origin to a maximum value, qv_p, where v_p is the maximum thermal quasi-particle group velocity; it resembles the continuous spectrum found for a normal system.

The above separation is directly analogous to that carried out in the previous chapter for the current-current response functions. It may be thought of as a generalization of the two-fluid concept to frequency and wave-vector dependent quantities. We may write:

$$\chi''(\mathbf{q}, \omega) = \chi''_s(\mathbf{q}, \omega) + \chi''_n(\mathbf{q}, \omega) \tag{7.1}$$

with $\chi''_s(\mathbf{q}, \omega)$, the superfluid component, denoting that part of χ'' arising from excitation of quasi-particles from the condensate. In the collisionless regime, the two components of χ'' will be distinct, the extent of their overlap being proportional to the (quite-small) probability that a thermal quasi-particle of momentum **q** decay into two quasi-particles of momentum $\mathbf{q} - \mathbf{p}$ and **p** as a result of scattering against the condensate. We shall show that these two components satisfy separate sum rules. Unlike the case at $T = 0$, sum rule considerations do not enable us to pin down precisely the quasi-particle sound velocity; they do, however, yield an order-of-magnitude estimate of the extent of its departure from the macroscopic, zero-temperature, value s.

In the hydrodynamic regime, one finds two modes of wave propagation, first and second sound. Superfluid hydrodynamics is richer than its Fermi liquid counterpart because one can have relative motion of the thermal quasi-particles and the condensate. In the first sound mode, the normal fluid and superfluid components move in phase with each other; it will be seen to be primarily a density wave, which resembles closely the hydrodynamic sound mode of a normal liquid. Second sound, on the other hand, corresponds to an out-of-phase motion of the two components; it is a wave motion characterized primarily by a periodic variation in the temperature of the system. One expects to see two peaks in $\chi''(\mathbf{q}, \omega)$ at the first and second sound frequencies; however, the amplitude of the second sound peak will be quite small, since second sound involves only a very slight fluctuation in the particle density. We shall see that sum rules permit us to fix the relative amplitudes of these two peaks.

We study first the propagation of first and second sound in the hydrodynamic limit and then go on to a consideration of quasi-particle sound in the collisionless regime. As usual, the transition from one regime to the other takes place for frequencies ω and relaxation times τ such that $\omega\tau \approx 1$. It corresponds here to a transition from first sound to quasi-particle sound, and will be characterized by a maximum in sound-wave attenuation. We discuss the experimental evidence for such a transition at the close of the chapter.

Our considerations will be confined to *reversible* phenomena. There are, of course, a wide variety of irreversible phenomena which have as their physical origin collisions between the thermally-excited quasi-particles. A most successful phenomenological account of such collisions, and their consequences, has been developed by Landau and Khalatnikov (1949). We refer the interested reader to Khalatnikov's book [Khalatnikov (1965)] for an account of their theory.

7.2 Two-Fluid Hydrodynamics: First and Second Sound

In order to derive the modes of propagation of first and second sound it is necessary that one consider the generalization of the two-fluid model to non-equilibrium situations, as set forth by Tisza and Landau. The microscopic basis for such a model resembles closely that used for the equilibrium situation. The normal fluid consists in the thermal quasi-particles; if these are in equilibrium in a frame of reference moving at a velocity $\mathbf{v}$ with respect to the condensate, they are described by a distribution function

$$n_p(\mathbf{v}) = \frac{1}{e^{\beta(\epsilon_p - \mathbf{p}\cdot\mathbf{v})} - 1} \tag{7.2}$$

This concept may be generalized to describe a non-equilibrium situation provided it is such that $\mathbf{v}$ and T are slowly-varying functions of space and time. Such variation should be sufficiently slow that one can still speak of *local* thermodynamic equilibrium, a condition which requires that spatial variations be slow compared to a mean free path, and temporal variations slow compared to the relaxation time needed to establish such local equilibrium. One considers as well variations in space and time of both the density and velocity of the condensate.

We shall here confine our attention to reversible fluid motion; we neglect the viscous effects which, as we have mentioned, necessarily accompany motion of the thermally-excited quasi-particles. The four basic equations which describe fluid flow then take a simple form: two conservation laws and two dynamic equations. The first conservation law is that of the local density of the system:

$$\frac{\partial \rho}{\partial t} = -\nabla \cdot \mathbf{J} \tag{7.3}$$

while the second is conservation of entropy. Let the entropy per unit volume of the system be $\tilde{S}$; since it is carried solely by the thermally-excited quasi-particles, the conservation law reads:

$$\frac{\partial \tilde{S}}{\partial t} = -\nabla \cdot \left(\tilde{S}\mathbf{v}_n\right) \tag{7.4}$$

The first dynamic equation governs condensate motion and is that we have derived in Chapter 5,

$$\frac{\partial \mathbf{v}_s}{\partial t} = -\nabla \left(\mu + \frac{1}{2}\rho \mathbf{v}_s^2\right) \tag{7.5}$$

The second dynamic equation governs motion of the entire liquid and is

$$\frac{d\mathbf{J}}{dt} = -\nabla P \tag{7.6}$$

where P is the local pressure acting on the liquid.

The three "new" equations, (7.3), (7.4), and (7.6) are nearly obvious; one can scarcely imagine them taking any other form. They may be derived by writing down the Boltzmann equation for the quasi-particle distribution function, and taking advantage of the conservation laws to eliminate the collision term. We refer the interested reader to, for example, the lectures of de Boer (1963), for the details of such a derivation.

It is illuminating to write the two dynamic equations in slightly different form; this may easily be done with the aid of the thermodynamic identity:

$$\rho d\mu = -\tilde{S}dT + dP \tag{7.7}$$

On making use of (7.7), and further, keeping only linear terms in the equations of motion, we find, on combining (7.5) and (7.6),

$$\rho_s \frac{\partial \mathbf{v}_s}{\partial t} = -\frac{\rho_s}{\rho}\nabla P + \frac{\rho_s}{\rho}\tilde{S}\nabla T \tag{7.8}$$

$$\rho_s \frac{\partial \mathbf{v}_n}{\partial t} = -\frac{\rho_n}{\rho}\nabla P - \frac{\rho_s}{\rho}\tilde{S}\nabla T \tag{7.9}$$

We see that a pressure gradient acts to drive both fluids in the same direction, while a temperature gradient acts to drive them in opposite directions. This latter aspect may be exhibited explicitly if one multiplies (7.5) by ρ and subtracts the resulting equation from (7.6): one finds, on using (7.7),

$$\rho_n \frac{\partial}{\partial t}(\mathbf{v}_n - \mathbf{v}_s) = -\tilde{S}\nabla T \tag{7.10}$$

A temperature gradient acts as an "osmotic" pressure, which tends to drive the fluids in opposite directions.

The above set of four basic equations may be reduced to two simple differential equations which govern the liquid motion. The first of these, the density equation of motion, is found by taking the time derivative of (7.3) and substituting on the right-hand side the appropriate result from (7.6); on keeping only linear terms, one finds thereby

$$\frac{\partial^2 \rho}{\partial t^2} = \nabla^2 P \tag{7.11}$$

The second, the entropy equation of motion, is obtained in similar fashion from (7.4), (7.9); making use of (7.11), we may write it as

$$\frac{\partial^2 \tilde{S}}{\partial t^2} = \frac{\tilde{S}^2}{\rho}\frac{\rho_s}{\rho_n}\nabla^2 T + \frac{\tilde{S}}{\rho}\frac{\partial^2 \rho}{\partial t^2} \tag{7.12}$$

This equation takes a yet simpler form, if one introduces the entropy per unit mass,

$$S = \frac{\tilde{S}}{\rho} \; ; \tag{7.13}$$

one then finds

$$\frac{\partial^2 S}{\partial t^2} = S^2 \frac{\rho_s}{\rho_n} \nabla^2 T \tag{7.14}$$

Periodic solutions for (7.11) and (7.14) are obtained by considering small departures of the pressure and temperature from equilibrium, according to

$$\delta P = \left(\frac{\partial P}{\partial \rho}\right)_S \delta\rho + \left(\frac{\partial P}{\partial S}\right)_\rho \delta S \tag{7.15a}$$

$$\delta T = \left(\frac{\partial T}{\partial \rho}\right)_S \delta\rho + \left(\frac{\partial T}{\partial S}\right)_\rho \delta S \tag{7.15b}$$

One searches, furthermore, for periodic solutions of the coupled equations, of the form:

$$\rho = \rho_o + \delta\rho \exp\left[i\left(\mathbf{q}\cdot\mathbf{r} - \omega t\right)\right] \tag{7.16a}$$

$$S = S_o + \delta S \exp\left[i\left(\mathbf{q}\cdot\mathbf{r} - \omega t\right)\right] \tag{7.16b}$$

The resulting density and entropy wave propagation is readily found if one neglects thermal expansion of the liquid, for then a change of pressure is not accompanied by a change in temperature, and, vice versa, a change in temperature is not accompanied by a change in density. Such an approximation is equivalent to assuming that $C_p = C_v$, and is quite accurate at the very low temperatures we consider. On making it, one sees at once that temperature and density waves are decoupled.

Indeed, on substituting equations (7.16) and (7.15) into (7.11) and (7.14), and continuing to keep only linear terms, we find the following dispersion relations for wave propagation:

Density Waves (First Sound)

$$\omega^2 = s_1^2 q^2 \; ; \quad s_1^2 = \left(\frac{\partial P}{\partial \rho}\right)_S \tag{7.17}$$

Entropy Waves (Second Sound)

$$\omega^2 = s_2^2 q^2 \; ; \quad s_2^2 = \frac{\rho_s}{\rho_n} S^2 \left(\frac{\partial T}{\partial S}\right)_\rho = \frac{\rho_s}{\rho_n}\left(\frac{S^2 T}{C_v}\right) \tag{7.18}$$

The velocity of the density wave is essentially that of the usual first sound wave in a normal liquid; the second sound velocity is seen to depend intimately on the presence of a superfluid component, and vanishes near the λ-point.

On referring back to (7.10), we see that in a first sound wave, the normal and superfluid components move in phase with each other: $\mathbf{v}_n = \mathbf{v}_s$. In a second sound wave, they move out of phase, in such a way that the net matter transport is negligible; according to (7.6),

$$\mathbf{J} = \rho_n \mathbf{v}_n + \rho_s \mathbf{v}_s = 0 \tag{7.19}$$

in a second sound wave.

It is not overly difficult to take into account the small coupling between density and temperature waves. One simply makes use of the complete equations, (7.15), and finds:

$$\left(\frac{\omega^2}{s_1^2 q^2} - 1\right)\delta\rho + \left(\frac{\partial P}{\partial S}\right)_\rho \left(\frac{\partial \rho}{\partial P}\right)_S \delta S = 0 \tag{7.20}$$

$$\left(\frac{\partial T}{\partial \rho}\right)_S \left(\frac{\partial S}{\partial T}\right)_\rho \delta\rho + \left(\frac{\omega^2}{s_2^2 q^2} - 1\right)\delta S = 0 \tag{7.21}$$

The condition that the equations be compatible yields a quadratic equation for ω^2, viz:

$$\left[\frac{\omega^2}{s_1^2 q^2} - 1\right]\left[\frac{\omega^2}{s_2^2 q^2} - 1\right] = \left[1 - \frac{C_v}{C_p}\right] \tag{7.22}$$

on making use of the appropriate thermodynamic identities. The right-hand side of (7.22) is, in fact, very small. It vanishes at $T = 0$, and is only 7×10^{-4} at $T = 1.5°\text{K}$ [London (1954)]. As a result, the two modes of wave propagation are effectively uncoupled, and their velocities accurately specified by (7.17) and (7.18).

The precise extent to which, for example, second sound involves a density fluctuation may be determined by simple sum rule considerations applied to the density fluctuation excitation spectrum. The relevant sum rules are those derived in Sec. 2.6, Vol. I, which for convenience, we reproduce here:

$$-\frac{1}{\pi}\int_{-\infty}^{+\infty} d\omega \chi''(\mathbf{q},\omega)\omega = \frac{Nq^2}{m} \tag{7.23}$$

$$\lim_{\mathbf{q}\to 0}\left\{-\frac{1}{\pi}\int_{-\infty}^{+\infty} d\omega \frac{\chi''(\mathbf{q},\omega)}{\omega}\right\} = \frac{N}{ms_i^2} \tag{7.24}$$

where s_i is the *isothermal* sound velocity. Let us assume that χ'' contains both a first and second sound peak, according to

$$\chi''(\mathbf{q},\omega) = -\frac{\pi N}{2m}\left\{\frac{Z_1}{s_1}\left[\delta\left(\omega - s_1 q\right) - \delta\left(\omega + s_1 q\right)\right]\right.$$
$$\left. + \frac{Z_2}{s_2}\left[\delta\left(\omega - s_2 q\right) - \delta\left(\omega + s_2 q\right)\right]\right\} \tag{7.25}$$

This expression is somewhat oversimplified, since both peaks will be broadened by viscous damping effects; however, as long as the peaks do not overlap (a situation well-satisfied in practice), the form, (7.25), suffices for considerations based on sum rules. On substituting (7.25) into the sum rules, and making use of the thermodynamic relation,

$$s_1^2 = \left(C_p/C_v\right) s_i^2 \tag{7.26}$$

one finds readily

$$Z_2 = \frac{C_p/C_v - 1}{s_1^2/s_2^2 - 1} \ll 1 \tag{7.27}$$

$$Z_1 = 1 - Z_2 \cong 1 \tag{7.28}$$

for the strength of the two poles in the density fluctuation spectral density. The result, (7.27), is in full accord with our general conclusion, based on (7.22), that the admixture of density fluctuation in a temperature wave will be of order $(C_p/C_v - 1)$.

Second sound is one of the most spectacular manifestations of the superfluid behavior of He II. Its propagation has been studied by many different experimental techniques. In Fig. 7.1 we plot the theoretical variation of s_2 with temperature. The experimental results obtained are in excellent agreement with the theoretical curve down to temperatures of the order of 0.7°K. Below this temperature, they begin to depart, for a very simple reason. At such temperatures the mean free path of a phonon is of the order of the size of the experimental apparatus, so that one no longer has sufficiently frequent collisions between the thermal excitations to establish the local thermodynamic equilibrium. (At $T = 0.8$°K, the calculations of Landau and Khalatnikov show that the phonon mean free path is of the order of 0.1 mm; it is greater than 1 cm at 0.5°K.) Under such circumstances, if one introduces a heat pulse at one end of the system (thereby creating phonons) a given excitation may simply propagate to the other end (at the quasi-particle sound velocity) without suffering any collisions. Experimentally one observes substantial distortion of

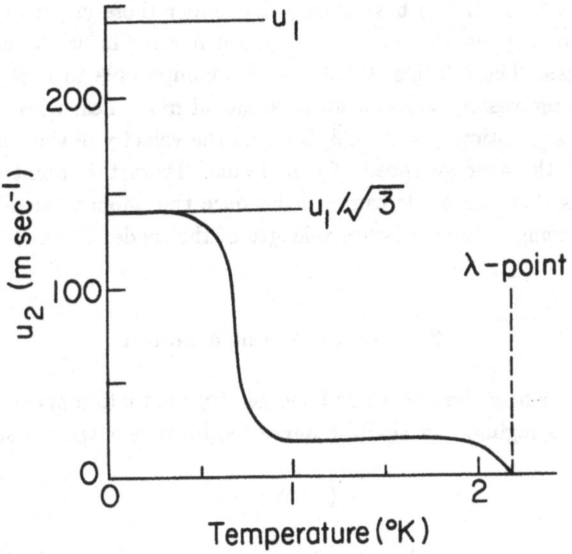

FIGURE 7.1. *The velocity of second sound at the vapor pressure [after Atkins (1959)].*

a temperature pulse of a sort which is consistent with such a physical picture.

We note from Fig. 7.1 that in the very low temperature region ($T \leq 0.5°K$), the theoretical value of the second sound velocity approaches a constant; one finds in fact

$$s_2 \simeq \frac{s_1}{\sqrt{3}} \quad (T \leq 0.5°\mathrm{K}) \tag{7.29}$$

This result may readily be found from (7.18); at such temperatures the only thermally-excited quasi-particles which are of importance are the phonons, whose velocity will be seen to be nearly s in this temperature region. It is straightforward to calculate ρ_n, S, and C_v for this phonon "gas"; it is left as an exercise for the reader to show that one thereby obtains (7.29).

Indeed, in this rather inaccessible region (from an experimental point of view, in view of the long phonon mean free paths), there exists an alternative derivation of the second sound mode. It is based on the fact that at such temperatures second sound corresponds to a change in density of the phonons (associated with the change in temperature). Ward and Wilks (1951, 1952) have shown that under those circumstances second sound may be viewed as a compression wave in an "almost ideal" phonon gas. The relation, (7.29), is then comparable to that which exists for compression waves in an ideal gas of molecules, where one finds the familiar relation, $c = \bar{u}_1/\sqrt{3}$, between the velocity of the compression wave and the average speed of a molecule. From this point of view it is obvious that the mode cannot exist once the phonon mean free path becomes comparable to the wave-length of the mode.

7.3 Quasi-Particle Sound

We consider now the behavior of the density fluctuation spectrum in the collisionless regime. We shall be interested in wave lengths λ such that

$$\xi \ll \lambda \ll \ell \tag{7.30}$$

where ξ is a coherence length (of the order of the interparticle spacing) and ℓ is the mean free path for the quasi-particle excitations under study. Where (7.30) is satisfied, one is both in the long wave-length regime and the collisionless regime. Under these circumstances, there will be two distinct contributions to $\chi''(\mathbf{q},\omega)$. We have identified these in (7.1) as a superfluid part, $\chi''_s(\mathbf{q},\omega)$, arising from excitation and de-excitation of single quasi-particles from the condensate, and a normal component, $\chi''_n(\mathbf{q},\omega)$, produced by the scattering of already thermally-excited quasi-particles.

The condition, (7.30), is not overly difficult to satisfy in practice. It is met, for example, in the neutron scattering experiments at $T = 1°$K for the lower range of q values studied (0.2 Å$^{-1} \leq q \lesssim 0.6$ Å^{-1}). Let us consider briefly the finite temperature analysis of such an experiment, which measures directly the dynamic form factor, $S(\mathbf{q},\omega)$.

As for $\chi''(\mathbf{q},\omega)$, there will be distinct contributions to $S(\mathbf{q},\omega)$, arising from transitions involving the condensate, and from scattering of the thermal quasi-particles. Since the latter contribution extends over a continuous range of frequencies, it is not easily separated from the background in an experiment; in practice, one measures only the "superfluid" component, $S_s(\mathbf{q},\omega)$, associated with the scattering of single quasi-particles in or out of the condensate. Inspection of (I.2.162) shows that we may write:

$$S_s(\mathbf{q},\omega) = NZ_q\left\{\delta\left(\omega-\omega_q\right)+n_q\left[\delta\left(\omega-\omega_q\right)+\delta\left(\omega+\omega_q\right)\right]\right\} \tag{7.31}$$

In (7.31) the first term in brackets corresponds to quasi-particle excitation; the second and third represent the induced excitation and de-excitation of single quasi-particles, respectively. The result, (7.31), is the natural finite-temperature analogue of the single quasi-particle part of $S(\mathbf{q}, \omega)$, (2.19). Both the quasi-particle energy, ω_q, and the transition probability, Z_q, depend on temperature, since the interaction of a given quasi-particle with the thermally-excited quasi-particles will be different from that it has with the condensate. We shall see that at 1°K, any departures of ω_q and Z_q from their zero-temperature values are too small to be picked up in a neutron scattering experiment in the collisionless regime. It is for this reason that we have taken the measurements of Henshaw, Woods, et al. at 1°K as a direct measure of the quasi-particle spectrum at $T = 0$. Let us emphasize that at wave vectors less than about 0.6 Å^{-1}, they have observed quasi-particle sound.

We note, too, that for temperatures T and wave-vectors q such that

$$\beta\omega_q \gg 1 \tag{7.32}$$

thermally-induced excitation and de-excitation of single quasi-particles is negligible. This condition is likewise met in the experiment of Henshaw and Woods at 1.1°K. We may thus justify, a posteriori, the use of a "zero-temperature" dynamic form factor, (2.19), in the analysis of their neutron scattering experiments.

We turn now to a consideration of the temperature dependence of the quasi-particle phonon energy, ω_q, or what is equivalent, the quasi-particle sound velocity, $s(T)$. We first establish two new sum rules for $\chi''(\mathbf{q}, \omega)$ in the collisionless regime; these permit us to obtain a qualitative measure of such temperature dependence. We then compare experimental measurements of $s(T)$ with theoretical calculations of this quantity in the very low temperature regime in which the only thermal excitations present are phonons.

To derive the sum rules we note first that current conservation provides the following relation between $\chi''_{\parallel}(\mathbf{q}, \omega)$, the imaginary part of the longitudinal current-current response function, and $\chi''(\mathbf{q}, \omega)$:

$$\chi''_{\parallel}(\mathbf{q}, \omega) = \frac{\omega^2}{q^2}\chi''(\mathbf{q}, \omega) \tag{7.33}$$

By analogy with (7.1), we can therefore separate $\chi''(\mathbf{q}, \omega)$ into a superfluid and a normal component, associated with condensate transitions and thermal quasi-particle transitions respectively. The components $\chi_{\parallel}^{n}$ and $\chi_{\parallel}^{s}$ obey the separate Kramers-Kronig relations:

$$\frac{1}{\pi}\int_{-\infty}^{+\infty} d\omega \frac{\chi_{\parallel}^{n''}(\mathbf{q},\omega)}{\omega} = \chi_{\parallel}^{n}(\mathbf{q},o) \tag{7.34a}$$

$$\frac{1}{\pi}\int_{-\infty}^{+\infty} d\omega \frac{\chi_{\parallel}^{s''}(\mathbf{q},\omega)}{\omega} = \chi_{\parallel}^{s}(\mathbf{q},o) \tag{7.34b}$$

Moreover, we have seen in the preceding chapter that in the long wave-length limit

$$\lim_{\mathbf{q}\to 0} \chi_{\parallel}^{n}(\mathbf{q},o) = \chi_{\perp}(\mathbf{q},o) = -\frac{\rho_n}{m^2} \tag{7.35a}$$

$$\lim_{\mathbf{q}\to 0} \chi_{\parallel}^{s}(\mathbf{q},o) = -\frac{\rho_s}{m^2} \tag{7.35b}$$

On making use of (7.33) and (7.35), we see that equation (7.34) provides directly the following sum rules:

$$\lim_{\mathbf{q}\to 0}\left[-\frac{1}{\pi}\int_{-\infty}^{+\infty} d\omega \chi_n''(\mathbf{q},\omega)\omega\right] = \frac{\rho_n q^2}{m^2} \tag{7.36}$$

$$\lim_{\mathbf{q}\to 0}\left[-\frac{1}{\pi}\int_{-\infty}^{+\infty} d\omega \chi_s''(\mathbf{q},\omega)\omega\right] = \frac{\rho_s q^2}{m^2} \tag{7.37}$$

The two sum rules represent an effective split-up of the f-sum rule, (7.23), in the long wave-length collisionless regime, into separate sum rules for χ_n'' and χ_s''. We have therefore only one *new* sum rule in this regime. It is natural to inquire for what values of q the various "long wave-length" sum rules, (7.24), (7.36), and (7.37) are valid. The condition for their validity is that there exist a *local* relation between the external field and the responding quantity. Thus the compressibility sum rule is valid as long as there exists a local relation between the external force field and the induced density; the superfluid sum rules are valid when one has a local relation between the induced current and the external "vector potential." Such local relations apply provided the wave-lengths of interest are long compared to the coherence lengths which characterize the system under study.

We now consider the information which the long wave-length sum rules provide on the quasi-particle sound velocity. The superfluid component, $\chi_s''(\mathbf{q},\omega)$, may be directly obtained from (7.31). We have

$$\chi_s''(\mathbf{q},\omega) = -\pi\left[S_s(\mathbf{q},\omega) - S_s(\mathbf{q},-\omega)\right]$$

$$= -\pi N Z_q \left[\delta\left(\omega - \omega_q\right) - \delta\left(\omega + \omega_q\right)\right] \tag{7.38}$$

On substituting (7.38) into (7.37), we find

$$\omega_q Z_q = \left(\frac{\rho_s}{\rho}\right)\left(\frac{q^2}{2m}\right) \tag{7.39}$$

A second relation between ω_q and Z_q is provided by the compressibility sum rule, (7.24). On substituting (7.38) and (7.1) into (7.24), we find:

$$\frac{Z_q}{\omega_q} = \frac{1}{2ms_i^2} + \lim_{q\to 0}\frac{1}{2\pi N}\int_{-\infty}^{+\infty} d\omega \frac{\chi_n''(\mathbf{q},\omega)}{\omega} \tag{7.40}$$

The two relations, (7.39), and (7.40) are not sufficient to determine ω_q, since we do not know $\chi_n''(\mathbf{q},\omega)$. We may, however, obtain a rough estimate of the integral in (7.40) by noting that the transitions contributing to χ_n'' involve the scattering of a thermal quasi-particle from state $\mathbf{p}$ to state $(\mathbf{p}+\mathbf{q})$. The corresponding excitation energy is equal to $\mathbf{q}\cdot\mathbf{v}_p$. We may thus write

$$\frac{\int_0^\infty \chi_n''(\mathbf{q},\omega)\omega d\omega}{\int_0^\infty \dfrac{\chi_n''(\mathbf{q},\omega)d\omega}{\omega}} = q^2\overline{v^2}$$

where $\overline{v^2}$ is of the order of the average squared group velocity of thermal quasi-particles. On making use of (7.36), we find

$$\lim_{q\to 0}\left[\frac{1}{\pi}\int_0^\infty \frac{\chi_n''(\mathbf{q},\omega)}{\omega}d\omega\right] = \frac{\rho_n}{m^2\overline{v^2}} = \frac{\rho_n}{\rho}A(T)\frac{N}{ms_i^2} \tag{7.41}$$

where we have set

$$A(T) = \frac{s_i^2}{\overline{v^2}} \tag{7.42}$$

$A(T)$ is a temperature dependent constant, which is of order unity, and which is subject to the inequality,

$$\frac{\rho_n}{\rho}A(T) < 1 \tag{7.43}$$

since the left-hand side of (7.40) is positive definite. On making use of (7.41), we obtain our desired result:

$$\omega_q = s_i q\left\{\frac{1-\dfrac{\rho_n}{\rho}}{1-\dfrac{A\rho_n}{\rho}}\right\}^{1/2} \tag{7.44}$$

An approximate microscopic calculation of the temperature-dependent quasi-particle sound velocity has been given by Hohenberg and Martin (1964). The present considerations permit one to set a lower limit on $s(T)$; according to (7.43), one has

$$s(T) > \left(\frac{\rho_s}{\rho}\right)^{1/2} s_i \tag{7.45}$$

Let us inquire whether such a temperature variation can be seen in neutron scattering experiments. We note that at 1.1°K, where the neutron scattering experiments have been carried out in the long wave-length regime, $\rho_n/\rho \sim 10^{-3}$. We therefore expect a shift from the zero temperature sound velocity which is of the order of 0.1%. Such an accuracy likely cannot be achieved in neutron experiments. However, to the extent that one can remain in the collisionless, long wave-length regime at higher temperatures, one may expect to see a measurable shift in the quasi-particle sound velocity from its zero-temperature value. Indeed, for temperatures near the λ-point, this sound velocity may tend toward zero, a temperature dependence which would be strikingly different from that observed (and anticipated) for the first sound velocity [Atkins (1959)].

The very slight dependence on temperature of the quasi-particle sound velocity in the low temperature regime (0.1°K to 0.8°K) has been observed by Whitney and Chase (1962), who use direct ultrasonic pulse experiments to measure the sound velocity at a frequency of 1 Mc. Their experimental results for the shift in the sound velocity from its zero-temperature value are shown in Fig. 7.2. Using Khalatnikov's calculations for the phonon relaxation times, one finds that $\omega\tau \sim 1$ for $T \lesssim 0.8$°K, so that the decrease observed in that vicinity may be attributed to the onset of hydrodynamic behavior, while the initial *increase* clearly takes place in the collisionless regime. The magnitude of the increase is not large ($\sim$ 1 cm/sec at $T = 0.4$°K); we now see to what extent it may be explained theoretically.

We may attempt an approximate calculation of $s(T)$ by calculating $\chi_n''(\mathbf{q},\omega)$ for a non-interacting gas of quasi-particles, and then using (7.41) to obtain $A(T)$. By starting with the exact expression for $\chi''(\mathbf{q},\omega)$, (I.2.165), and following steps directly analogous to those used to derive (6.22), one finds

$$\chi_n''(\mathbf{q},\omega) = -\pi \sum_{\mathbf{q}} \left(n_{\mathbf{p}} - n_{\mathbf{p+q}}\right) F_{\mathbf{pq}}^2 \delta\left(\omega - \varepsilon_{\mathbf{q+p}} + \varepsilon_{\mathbf{p}}\right) \tag{7.46}$$

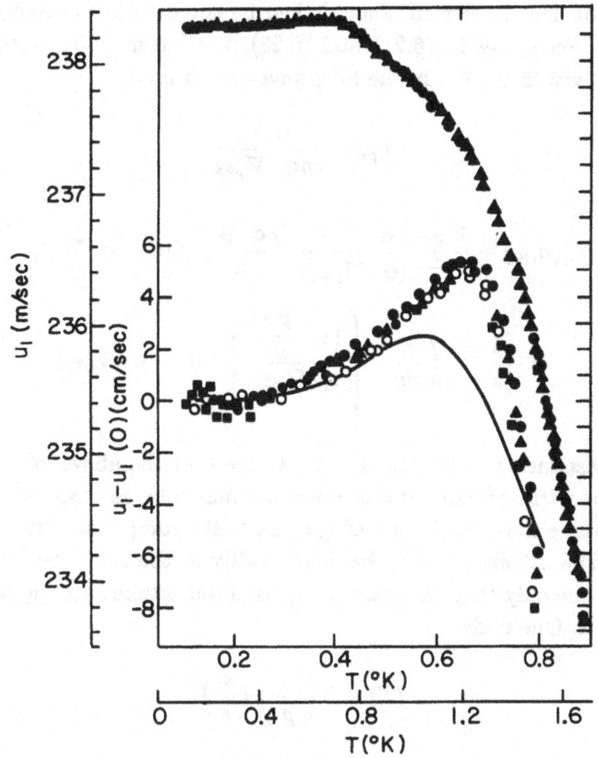

FIGURE 7.2. *Velocity of sound in liquid helium II. The lower curve shows the region of the maximum on an expanded scale [from Whitney and Chase (1962)].*

where $F_{\mathbf{pq}}$ is the matrix element for scattering of a quasi-particle from a state $\mathbf{p}$ to a state $\mathbf{p}+\mathbf{q}$,

$$F_{\mathbf{pq}} = \langle \mathbf{p}+\mathbf{q}| \sum_{\mathbf{p}'} C^{+}_{\mathbf{p}'+\mathbf{q}} C_{\mathbf{p}'} |\mathbf{p}\rangle \tag{7.47}$$

Current conservation provides a relation between $F_{\mathbf{pq}}$ and the corresponding matrix element for the longitudinal current density fluctuation,

$$(\varepsilon_{\mathbf{p}+\mathbf{q}} - \varepsilon_{\mathbf{p}})\, F_{\mathbf{pq}} = \langle \mathbf{p}+\mathbf{q}| \sum_{\mathbf{p}'} C^{+}_{\mathbf{p}'+\mathbf{q}} C_{\mathbf{p}'} \left(\mathbf{p}' \cdot \frac{\mathbf{q}}{m}\right) |\mathbf{p}\rangle \tag{7.48}$$

We now assume that the current fluctuations are symmetric (as would be the case for non-interacting quasi-particles); the longitudinal current fluctuation matrix element should then be equal to its transverse counterpart. According to (6.21) and (6.25), the latter is $(\mathbf{p}\cdot\eta_q/m)$. We therefore would expect, in the long wave-length limit,

$$F_{\mathbf{pq}} = \frac{\mathbf{q}\cdot\mathbf{p}}{m\mathbf{q}\cdot\nabla_p\varepsilon_p} \tag{7.49}$$

and

$$\chi''_n(\mathbf{q},\omega) - \frac{1}{\pi}\sum_{\mathbf{p}}\frac{\mathbf{q}\cdot\nabla_{\mathbf{p}}n_p}{(\mathbf{q}\cdot\nabla_{\mathbf{p}}\varepsilon_p)^2}\left(\frac{\mathbf{q}\cdot\mathbf{p}}{m}\right)^2\delta\left(\omega-\mathbf{q}\cdot\nabla_{\mathbf{p}}\varepsilon_p\right)$$

$$= -\frac{1}{\pi}\sum_{\mathbf{p}}\left(\frac{\partial n}{\partial\varepsilon}\right)\left\{\frac{\left(\mathbf{q}\cdot\frac{\mathbf{p}}{m}\right)^2}{\mathbf{q}\cdot\nabla_{\mathbf{p}}\varepsilon_p}\right\}\delta\left(\omega-\mathbf{q}\cdot\nabla_{\mathbf{p}}\varepsilon_p\right) \tag{7.50}$$

On substituting (7.50) into (7.36), we see that the above expression is consistent with the normal quasi-particle sum rule, as it should be. We next evaluate the contribution of $\chi''_n(q,\omega)$ to the compressibility sum rule, (7.41). The calculation may be done readily in the temperature regime in which the only thermal quasi-particles of importance are the phonons, $T \lesssim 0.5$°K. One finds

$$A(T) = 3\left(\frac{\rho_n}{\rho}\right)\left(\frac{s_i}{s}\right)^2 \tag{7.51}$$

If we now substitute (7.51) into (7.43), and keep only the lowest order terms in ρ_n/ρ, we obtain

$$s(T) = s_i(T)\left\{1+\frac{\rho_n}{\rho}\right\} \tag{7.52}$$

The non-interacting excitation gas calculation of $s(T)$ thus predicts a modest increase over the isothermal sound velocity s_i; the origin of the increase lies in the fact that at these temperatures the thermal quasi-particle contribution to the compressibility sum rule increases more rapidly with increasing temperature than does their contribution to the f-sum rule. At higher temperatures this is likely not the case.

The shift (7.52) has the right sign to explain the measurements of Whitney and Chase; however, its magnitude is far too small, since δs at 0.4°K is of the order of 1 cm/sec. Coherence effects in the scattering of thermally-excited phonons thus must play an important role. A calculation in which such effects are taken into account has been carried out by Andreev and

Khalatnikov (1963); we refer the interested reader to their paper for the details of the calculation. They find (in c.g.s. units)

$$s(T) = s(0) + 20\,T^4 \ln\left(\frac{67}{T^2}\right)\ (T \lesssim 0.5°\text{K}) \tag{7.53}$$

The temperature variation implied by (7.53) is considerably more rapid than that found using (7.52). The agreement with experiment is consequently better, though not perfect, as may be seen in Fig. 7.2.

7.4 Transition from Quasi-Particle Sound to First Sound

We have emphasized that in the vicinity of $\omega\tau \sim 1$, one gets a transition from quasi-particle sound to first sound. If one works at fixed frequency and increases the temperature, one expects to find quasi-particle sound at the lowest temperatures, then a rather complicated transition region, followed by first sound in the "high temperature" region for which $\omega\tau \ll 1$. We have just seen that the velocity of "sound" measured by Whitney and Chase, shows a maximum at a temperature such that $\omega\tau \sim 1$. Below that temperature ($\sim 0.75°$K), one finds quasi-particle sound with a velocity which increases with temperature: above it, the hydrodynamic sound velocity decreases with increasing temperature, in accordance with theoretical expectations.

A rather more striking manifestation of the transition region is found in ultrasonic attenuation experiments. In the very low temperature, quasi-particle sound regime, the sound wave attenuation will simply be proportional to $1/\tau$, the lifetime of a given phonon at that frequency. On the other hand, at "high" temperatures where one is in the hydrodynamic regime, the sound wave attenuation is determined by the appropriate viscosity coefficient. The corresponding attenuation is reduced over that found in the collisionless regime by a factor of $(\omega\tau)^2$, and is therefore proportional to τ. As in the case of the Fermi liquid, the general behavior of the ultrasonic attenuation coefficient with temperature should therefore be governed by an expression of the form

$$\frac{A}{\tau}\left\{\frac{\omega^2\tau^2}{1+\omega^2\tau^2}\right\} \tag{7.54}$$

According to (7.54), one expects to find a maximum in the ultrasonic attenuation coefficient in the region $\omega\tau \sim 1$. This is exactly what is observed, as may be seen from the results of Chase and Herlin at 12.1 Mc,

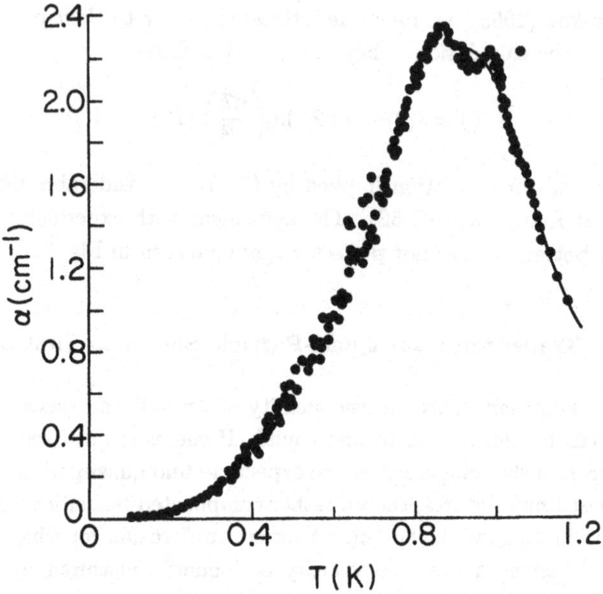

FIGURE 7.3. *Attenuation of sound in liquid helium at 12.1 Mc/sec [from Chase and Herlin (1955)].*

shown in Fig. 7.3. Above 0.8°K, the observed attenuation is in good agreement with the theory of Khalatnikov, whose results are shown there. Below 0.8°K, agreement between theory and experiment is not satisfactory, essentially since new physical processes come into play in the attenuation of quasi-particle sound at these lower temperatures. Let us emphasize that we lack a precise treatment of the transition regime, since it comprises just that regime for which neither first-sound nor quasi-particle sound offer an altogether satisfactory description of the system behavior.

CHAPTER 8

VORTEX LINES

8.1 Structure of a Vortex Line

In the preceding chapters, we have discussed at length the longitudinal oscillations of a superfluid Bose liquid. We now focus our attention on "steady" superfluid flow, in which the superfluid velocity, $\mathbf{v}_s$, satisfies the equation

$$div\ \mathbf{v}_s = 0$$

Such flow can, in principle, occur only in multiply-connected systems. However, it is possible to achieve an equivalent situation in the bulk of the liquid by setting up vortex lines. In a *vortex line* γ, the fluid rotates around the curve γ; the flow is irrotational everywhere *except on* γ. As a consequence, the velocity increases as one approaches γ. At a small enough distance, the centrifugal force is (in principle) large enough to overcome the capillary force, so that one expects to find a narrow cylindrical hole in the liquid along the vortex line. The existence of the hole makes the system multiply connected. Actually, such a picture is physically false, as the computed radius of the expected hole is comparable to the interparticle spacing: on such a small scale, the concept of a fluid flow is meaningless. We should therefore consider a vortex line as an irrotational motion of the fluid around the line γ, down to distances at which the hydrodynamic equations are no longer valid.

It should be realized that vortex lines are not peculiar to superfluids. They constitute as essential feature of the dynamics of normal fluids, and have, indeed, been extensively studied since the 19th century. Many of the results which we shall discuss are actually "classics" of hydrodynamics.

The major new feature brought in by superfluidity, and a most important one indeed, is the quantization of vortex lines, arising as a consequence of the quantization of circulation. In cases where quantization may be neglected, the properties of vortex lines in a superfluid are very close to those in a normal fluid.

We consider first a straight vortex line, in which the fluid rotates around the z-axis. The superfluid velocity at an arbitrary point M is tangential, as shown in Fig. 8.1; it depends only on the distance r between M and the axis of the vortex. In order for the flow to be irrotational, the magnitude of the velocity must vary as

$$v_s = \frac{k}{r} \tag{8.1}$$

where k is a constant. Equation (8.1) is valid for large distances. At small distances, when r becomes comparable to the coherence length, ξ, $\mathbf{v}_s$ varies rapidly, and the hydrodynamic description breaks down. We may thus consider that the vortex possesses a core of radius ξ, inside which the microscopic structure of the fluid is appreciably altered. For liquid He II, this core is of the order of a few Å wide.

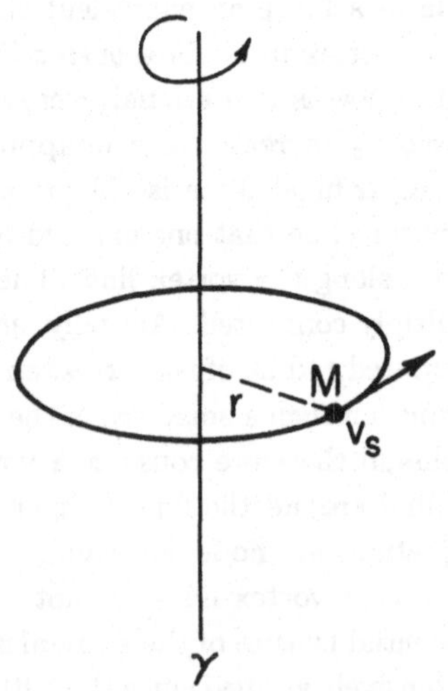

FIGURE 8.1. *Geometry of a straight vortex line.*

The circulation of the velocity (8.1) around a circle of radius r is equal to $2\pi k$. On taking account of the quantization of circulation, (5.50), we may write (8.1) as

$$v_s = \frac{n\hbar}{mr} \tag{8.2}$$

where n is an integer, giving the number of quanta of circulation in the vortex line (the sign of n determines the direction of the flow). We shall see that at low enough temperatures, only vortex lines with $|n| = 1$ are excited.

The energy per unit length of the vortex line, E, may be interpreted as a "tension" of the line. The kinetic energy associated with the rotation is given by

$$\int_o^R 2\pi r dr \frac{Nm}{2} v_s^2 = \frac{\pi N}{m} n^2 \hbar^2 \int_o^R \frac{dr}{r} \tag{8.3}$$

where R is the radius of the vessel containing the fluid. The integral, (8.3), diverges for small r; however, there exists a natural cutoff at the coherence length, ξ. With logarithmic accuracy, we can thus write

$$E = \frac{\pi N}{m} n^2 \hbar^2 \log \frac{R}{\xi} \tag{8.4}$$

In principle, we should add to (8.4) the change in potential energy brought about by the vortex. Actually, it may be shown that this correction affects only the immediate vicinity of the core ($r \sim \xi$); the corresponding contribution to the energy is comparable to the uncertainty inherent to the logarithmic accuracy of (8.4), and may thus be neglected.

To the extent that the finite size of the vortex core may be neglected, the velocity field $\mathbf{v}_s$ given by (8.2) satisfies the relation

$$\text{curl } \mathbf{v}_s = C\eta\delta_2(\mathbf{r}) \tag{8.5}$$

where η is a unit vector along the vortex axis, while

$$C = \frac{nh}{m} \tag{8.6}$$

is the circulation around the vortex line. $\delta_2(\mathbf{r})$ is a two-dimensional δ-function in the plane perpendicular to the vortex line. Equation (8.5) may be extended to describe the properties of curved vortices, such as vortex rings (a vortex line closed upon itself). In that more general case, $\delta_2(\mathbf{r})$ is defined at every point M of the vortex line as a δ-function in the plane normal to the vortex; η is a unit vector tangent to the vortex line

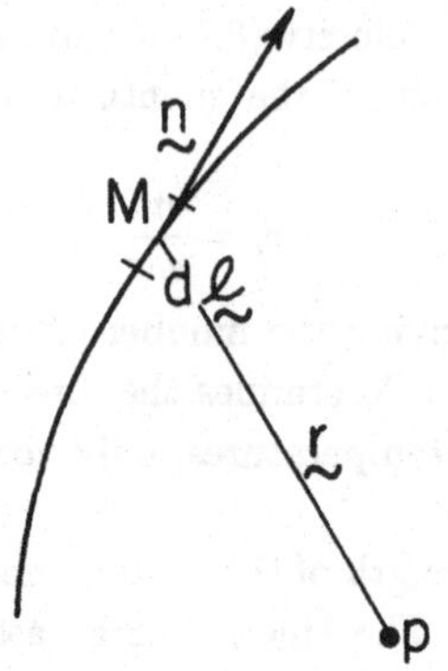

FIGURE 8.2. *Geometry of a curved vortex line.*

at M (Fig. 8.2). The velocity $\mathbf{v}_s$ created by a curved vortex at a point P cannot generally be obtained in closed form; however, it may be written as a line integral along the vortex:

$$\mathbf{v}_s = \frac{C}{4\pi} \int_\gamma \frac{\eta \times \mathbf{r}}{r^3} d\ell \tag{8.7}$$

(where $\mathbf{r}$ is the vector going from M to P). The expression (8.7) is formally similar to that giving the magnetic field created by a current carrying wire; its proof is left as an exercise to the reader. The velocity pattern of such curved vortices is complicated; at a distance r which is small compared to the radius of the curvature of γ, $\mathbf{v}_s$ may be written as

$$\mathbf{v}_s = \frac{C}{2\pi r^2} \eta \times \mathbf{r} + \bar{\mathbf{v}} \tag{8.8}$$

The first term on the right-hand side of (8.8) arises from that part of the vortex which lies in the immediate vicinity of the point P under study. The other, regular, contribution $\bar{\mathbf{v}}$ describes the velocity created at point P by the rest of the vortex.

The analysis is easily extended to the case of several vortices. The total velocity $\mathbf{v}_s$ is simply the sum of the velocities arising from each vortex. Let us for instance consider two parallel straight vortex lines γ_1 and γ_2, at a distance d from each other, possessing respectively n_1 and n_2 quanta of circulation. The net velocity is

$$\mathbf{v}_s = \mathbf{v}_1 + \mathbf{v}_2 \; ; \tag{8.9}$$

the total kinetic energy per unit length along the vortices is

$$\frac{Nm}{2}\iint d\sigma\,(v_1+v_2)^2 \tag{8.10}$$

The square terms in (8.10) correspond to the energy of the single vortex lines, while the cross term describes an *interaction energy* between two vortex lines. After a straightforward integration, one finds that the interaction energy is equal to

$$E_{12} = 2\pi\frac{N}{m}n_1 n_2 \hbar^2 \log\frac{R}{d} \tag{8.11}$$

where R is the radius of the vessel. [(8.11) is valid only if the two vortex lines in question are not too close to the vessel boundary.]

The total energy of the two vortex lines is equal to

$$E = \frac{\pi N}{m}\hbar^2\left\{\left(n_1^2+n_2^2\right)\log\frac{R}{\xi} + 2n_1 n_2 \log\frac{R}{d}\right\} \tag{8.12}$$

which we write in the form

$$E = \frac{\pi N}{m}\hbar^2\left\{\left(n_1+n_2\right)^2\log\frac{R}{\xi} - 2n_1 n_2 \log\frac{d}{\xi}\right\} \tag{8.13}$$

For a given value of the *total* circulation, one thus obtains a lower energy by having two vortices, with n_1 and n_2 circulation quanta respectively, in place of a single one with (n_1+n_2) quanta. Consequently, the lowest energy will be achieved by having a large number of vortex lines, each with a single quantum ($n=1$), and no vortices of higher order. At $T=0$, we thus expect to observe only "elementary" vortex lines, containing a single circulation quantum.

8.2 Dynamics of a Vortex Line

The motion of a vortex line is governed by a very simple law: each point M of the vortex moves at the velocity which the fluid possesses at M itself. More exactly, in the vicinity of M, the velocity $\mathbf{v}_s$ may be written in the form (8.8): the point M then moves at velocity $\bar{\mathbf{v}}$. Consider, for example, the two parallel vortices described earlier; each moves at the velocity created at its core by the other. If $n_1 = n_2$, the two vortices rotate around each other (Fig. 8.3a); if, instead, $n_1 = -n_2$, the vortices are subject to a uniform translation (Fig. 8.3b). Another example of such

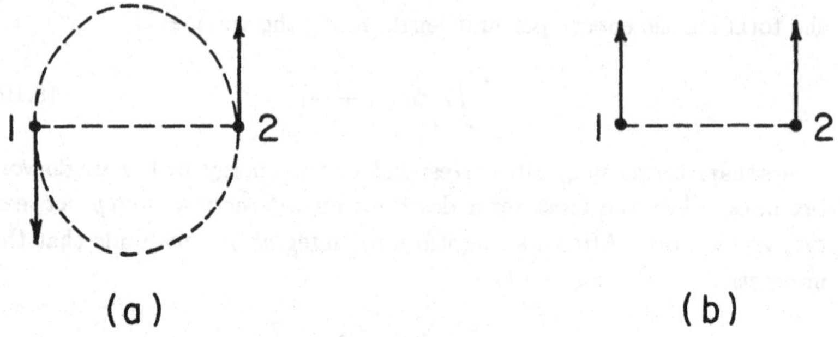

FIGURE 8.3. *Relative motion of two parallel vortices: (a) With equal circulation; (b) With opposite circulation.*

motion is the circular vortex ring (Fig. 8.4). At every point M, the rest of the vortex creates a velocity $\bar{\mathbf{v}}$ which is parallel to the ring axis (for reasons of symmetry); the magnitude of $\bar{\mathbf{v}}$ depends on the radius of the ring, r, and is found to be[1]

$$|\bar{\mathbf{v}}| \simeq \frac{\hbar}{2mr} \log \frac{r}{\xi} \tag{8.14}$$

The vortex ring moves along its axis at the *constant* velocity $\bar{\mathbf{v}}$ (as does for instance a smoke ring). Such motion is observed directly in the experiments of Rayfield and Reif (1963) on charged vortex rings.

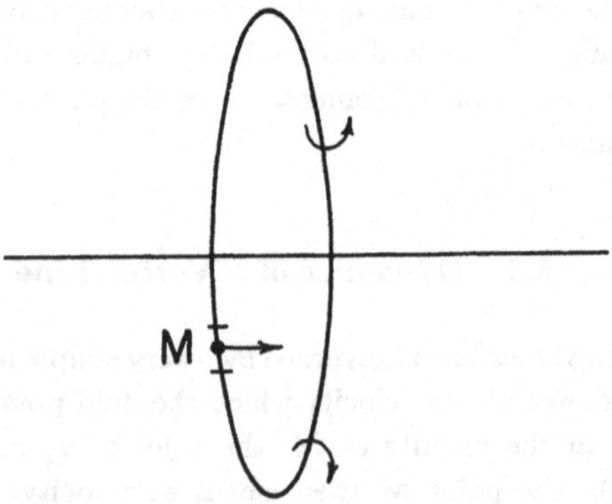

FIGURE 8.4. *Motion of a circular vortex ring.*

[1]See, for instance, H. Lamb (1945).

In order to prove the fundamental law of vortex motion, we start from equation (5.86), describing the dynamic features of superfluid flow:

$$Nm\frac{d\mathbf{v}_s}{dt} = Nm\left\{\frac{\partial \mathbf{v}_s}{\partial t} + (\mathbf{v}_s \cdot \mathbf{grad})\,\mathbf{v}_s\right\} = -\mathbf{grad}\ \mu \qquad (8.15)$$

(where μ is the chemical potential). (8.15) may also be written as

$$\frac{\partial \mathbf{v}_s}{\partial t} + \frac{1}{2}\mathbf{grad}\ \mathbf{v}_s^2 - \mathbf{v}_s \times \mathbf{curl}\ \mathbf{v}_s = -\frac{1}{Nm}\mathbf{grad}\ \mu \qquad (8.16)$$

On taking the curl of (8.16), we find the well-known equation describing the transport of vorticity

$$\frac{\partial \mathbf{u}}{\partial t} = \mathbf{curl}\ (\mathbf{v}_s \times \mathbf{u}) \qquad (8.17)$$

where we have set

$$\mathbf{u} = \mathbf{curl}\ \mathbf{v}_s \qquad (8.18)$$

We first consider the left-hand side of (8.17). $\mathbf{u}$ is localized at the core of the vortex. If the point M moves at velocity $\mathbf{v}_L$ (which represents the *vortex line velocity*), it follows that

$$\frac{\partial \mathbf{u}}{\partial t} = -\,(\mathbf{v}_L \cdot \mathbf{grad})\,\mathbf{u} \qquad (8.19)$$

On the right-hand side of (8.17) we replace $\mathbf{v}_s$ by its expression (8.8). The first, divergent, term of $\mathbf{v}_s$ gives a cross product which is *radial* in symmetry, and thus possesses no curl. If we further remark that $\bar{\mathbf{v}}$ may be considered as constant in the region where u is important, we may cast (8.17) in the form,

$$-\,(\mathbf{v}_L \cdot \mathbf{grad})\,\mathbf{u} = -\,(\bar{\mathbf{v}} \cdot \mathbf{grad})\,\mathbf{u} \qquad (8.20)$$

It follows that

$$\mathbf{v}_L = \bar{\mathbf{v}} \qquad (8.21)$$

which proves that the vortex line moves with the fluid velocity $\bar{\mathbf{v}}$ created at its core.

The relation (8.21) suffices to describe the motion of one or several vortices; there is no need for further dynamical laws. However, it is sometimes found convenient to express this law of motion in another, equivalent way, involving an inertial force known as the Magnus force. Such a point of

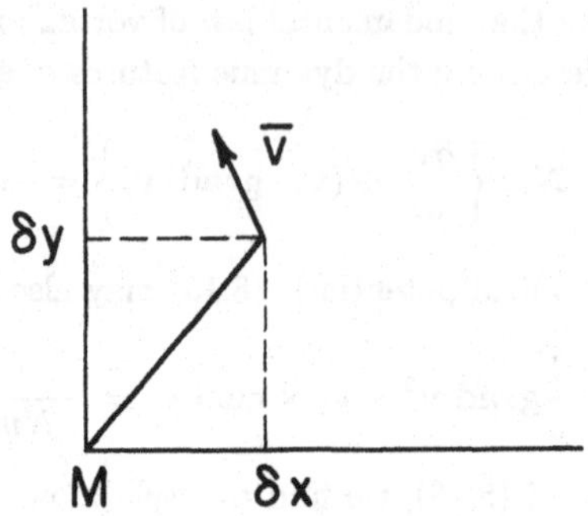

FIGURE 8.5. *Transverse displacement of an oscillating vortex line.*

view clearly displays the gyroscopic nature of vortex motion. Since it is completely equivalent to the law (8.21), we shall not describe it in detail. The interested reader is referred to standard text books in fluid dynamics.

As an example of such vortex motion, let us consider the *oscillations* of a single vortex line. Such oscillations were studied long ago by Thomson for a classical fluid; we shall only sketch their physical origin. We consider a straight vortex line parallel to the z-axis, which we stretch into a helix. As shown in Fig. 8.5, each point M of the vortex line is displaced in the x and y directions by the respective amounts

$$\delta x = a \cos kz$$

$$\delta y = a \sin kz \tag{8.22}$$

As a result of this stretching, the velocity pattern is altered. At every point M of the vortex, the velocity $\mathbf{v}_s$ takes the form (8.8). Detailed calculation shows that $\bar{\mathbf{v}}$ is perpendicular to the vortex displacement, with a modulus

$$|\bar{\mathbf{v}}| = a\frac{C}{4\pi}k^2 \log\left(\frac{1}{k\xi}\right) \tag{8.23}$$

According to the general law of vortex motion, the vortex moves at velocity $\bar{\mathbf{v}}$. It thus performs a circular motion around the z-axis, at a frequency

$$\omega = \frac{|\bar{\mathbf{v}}|}{a} = \frac{C}{4\pi}k^2 \log\left(\frac{1}{k\xi}\right) = \frac{\hbar k^2}{2m}\log\left(\frac{1}{k\xi}\right) \tag{8.24}$$

(assuming that the vortex carries one quantum of circulation). The helical deformation of the vortex thus *propagates*, giving rise to a circularly polarized oscillation. We note the non-analytic behavior of ω as a function of k. It may be verified that the oscillating vortex rotates in a direction opposite to that of the fluid rotation inside the vortex.

Such oscillations represent a new branch in the elementary excitation spectrum of a Bose liquid, which exists only in the presence of vortices. They have been observed directly by Hall and Vinen (1956) who were able to set up standing vortex waves in a stack of oscillating discs in liquid helium. The reader is referred to their article for details of the experiment.

8.3 Interaction Between Vortex Lines and the Normal Fluid: Mutual Friction

In the absence of vortex lines, the superfluid and normal components of a Bose liquid are completely decoupled. There is no mechanism by which superfluid motion can transfer energy to the normal fluid, thereby dissipating heat. This is no longer true in the presence of vortices. The latter provide an unexpected coupling between the superfluid and the thermal excitations, which may be regarded as a *mutual friction* between the two fluids. Such friction leads to heat dissipation; it also acts to damp the vortex oscillations discussed in the preceding section. The theory of mutual friction is still incomplete. We shall only attempt to stress its salient features.

We first consider a rigid vortex line, moving as a whole through the fluid. Because it represents an *inhomogeneity* in the superfluid structure it acts to scatter normal quasi-particles. The scattering gives rise to a momentum transfer between the normal fluid and the vortex line, which on the average leads to an *interaction force.* For a fixed vortex in a normal fluid at rest, the scattering from state $\mathbf{p}$ to $\mathbf{p}'$ is balanced by that from $\mathbf{p}'$ to $\mathbf{p}$: the net momentum transfer vanishes by symmetry. If, however, there is a relative motion of the vortex (with velocity $\mathbf{v}_L$) with respect to the normal fluid (with velocity $\mathbf{v}_n$), the net momentum transfer is no longer zero. Up to first order in $(\mathbf{v}_L - \mathbf{v}_n)$, the corresponding interaction force may be written as

$$\mathbf{F} = \alpha\,(\mathbf{v}_L - \mathbf{v}_n) + \beta\eta \times (\mathbf{v}_L - \mathbf{v}_n) \tag{8.25}$$

(where η is a unit vector along the vortex core). The first term on the right-hand side of (8.25) is a friction force; the work associated with it provides the energy dissipated into heat. The second term on the right-hand side of (8.25) is perpendicular to the relative motion, and thus does no work. It arises because of the asymmetry produced by the fluid rotation around the vortex line.

The coefficients α and β may be calculated in the framework of a semi-classical treatment in which a quasi-particle with momentum $\mathbf{p}$, embedded in a superfluid flow with velocity $\mathbf{v}_s$, is assumed to have the energy

$$\varepsilon_p + \mathbf{p} \cdot \mathbf{v}_s$$

The velocity pattern of the vortex gives rise to a position-dependent energy, which may be used to calculate the scattering. The cross-section was first evaluated in the Born approximation by Hall and Vinen; an improved calculation, based on the WKB approximation, has been carried out by Ginzburg and Pitaevskii (1958). Both calculations neglect the (possibly important) scattering by potential fluctuations at the vortex core.

In applying (8.25) one must also take account of the fact that the normal fluid flow is distorted by its coupling to the vortex. What should enter (8.25) is the normal velocity $\mathbf{v}_n$ at the vortex core as distorted by mutual friction, rather than the velocity $\mathbf{v}_{no}$ away from the vortex. The relation between $\mathbf{v}_n$ and $\mathbf{v}_{no}$ is controlled by the viscosity of the normal fluid (in fact, it is through viscous forces that energy is ultimately dissipated into heat). We refer the reader to the review article by Vinen (1961a) for a detailed discussion of this point.

As a result of the force (8.25), the vortex no longer moves at the "external" superfluid velocity $\mathbf{v}_s$ at its core. The difference $(\mathbf{v}_L - \mathbf{v}_s)$ may be calculated by balancing the friction force (8.25) with the Magnus force. Another approach is to show that the velocity pattern of a vortex line is modified by its motion relative to the normal fluid. In addition to the usual term (8.2), $\mathbf{v}_s$ contains a correction $\bar{\mathbf{v}}$ which incorporates the effect of mutual friction. According to (8.21), the core moves at the *total* superfluid velocity:

$$\mathbf{v}_L = \mathbf{v}_s + \bar{\mathbf{v}} \tag{8.26}$$

(8.26) is identical with the result obtained by balancing the Mangus force against the friction force.

The dynamics of superfluid flow is thus appreciably modified by the presence of vortex lines. For instance, the vortex oscillation studied in the preceding section is in fact damped; the vortex line spirals inward until it has reached its equilibrium position; the energy of the oscillation is dissipated into heat. Another example is the propagation of second sound in a direction perpendicular to the vortex. We have seen that second sound is nothing but a motion of the superfluid relative to the normal fluid. The vortex, which tends to follow the superfluid part, is slowed down by friction on the normal component. The corresponding dissipation results in attenuation of the second sound wave. Such a mechanism does not exist for second sound propagating parallel to the vortex, since in that case the vortex does not move. We thus expect to find a large anisotropy in the attenuation of second sound when vortices are present. Such an effect, which has been observed by Hall and Vinen, represented the first direct experimental evidence for the existence of vortex lines.

The existence of mutual friction has received ample experimental confirmation. On the other hand, the corresponding theory is not completely satisfactory. For instance, we have ignored the dynamic coupling of the vortex line to the normal fluid. When a thermal quasi-particle scatters against the vortex, it may *excite* the vortex oscillations. In other words, the vortex is not rigid, and may respond to quasi-particle scattering. Such "inelastic" processes may change considerably the magnitude of the mutual friction force. Also, the linear approximation (8.25) has a limited range of validity. Experimentally, non-linear effects are found to occur at rather small values of $(\mathbf{v}_L - \mathbf{v}_n)$. Such difficulties are hardly surprising, since they already occur in the description of vortex motion in normal fluids.

8.4 Vortex Lines in a Rotating Bucket[2]

We now proceed to apply the foregoing considerations to the practical problem discussed in Chapter 4, namely, the rotating bucket experiment. We have seen that the equilibrium configuration was obtained by minimizing the energy $(E - \mathbf{L} \cdot \omega)$ in the rotating frame of reference. To the extent that the condensed phase is fixed, we found that the equilibrium corresponded to $L = 0$; the fluid should thus be at rest, even though the bucket is rotating. As we have remarked, this is disproved by experiment. The surface of the liquid has the usual meniscus shape, which indicates that, on the average, the fluid rotates along with the bucket. The motion arises through the appearance of vortex lines which are parallel to the rotation axis in the rotating fluid.

[2]The treatment in this section follows closely that given by Feynman (1955).

To see how this comes about, let us minimize the quantity $(E - \mathbf{L} \cdot \omega)$ for a state containing ν vortex lines per unit surface perpendicular to the z-axis. In order to have as small a kinetic energy as possible, we must only use vortex lines with a single circulation quantum ($n = 1$), and set them in a roughly equidistant array (so that their interaction energy is a minimum). We thus expect the vortex line density ν to be constant over the whole section of the bucket.

The velocity at point $\mathbf{r}$ arising from a vortex line centered at point $\mathbf{r}_o$ may be conveniently written as

$$\mathbf{v}(\mathbf{r}) = \frac{\hbar}{m} \text{curl } \{\eta \log |\mathbf{r}_o - \mathbf{r}|\} \tag{8.27}$$

where η is a unit vector in the z-direction. The total velocity is obtained by summing (8.27) over all the vortex lines present in the fluid. Let us assume that there are *many* vortex lines. In order to calculate the average velocity over a distance which is large compared to the spacing between vortices, we replace the summation by an integration. We thus find

$$\langle \mathbf{v}(\mathbf{r}) \rangle = \frac{\nu\hbar}{m} \text{curl } \left\{ \eta \iint d\sigma_o \log |\mathbf{r} - \mathbf{r}_o| \right\} \tag{8.28}$$

The integration is easily performed for a cylindrical bucket with radius R. One obtains

$$\langle \mathbf{v}(\mathbf{r}) \rangle = \frac{\pi\nu\hbar}{m} \text{curl } \eta \times \mathbf{r} \tag{8.29}$$

The average velocity created by a large number of evenly distributed vortex lines thus corresponds to a uniform rotation around the z-axis, with an angular velocity

$$\omega_o = \frac{\pi\nu\hbar}{m} \tag{8.30}$$

Although the *local* motion of the fluid is irrotational, the *average* velocity field is not. The curl of $\langle \mathbf{v}(\mathbf{r}) \rangle$ is instead equal to the circulation per unit area, i.e., to $\nu 2\pi\hbar/m = 2\omega_o$, in agreement with (8.29).

Over distances smaller than the average spacing between vortex lines, the velocity $\mathbf{v}(\mathbf{r})$ fluctuates, displaying a sharp singularity at the core of every vortex line. Such a velocity pattern is schematically pictured on Fig. 8.6 (where 2ξ is the diameter of the *core*). Such discontinuous behavior arises as a consequence of the *quantization* of vortex lines in a superfluid. In a normal fluid one may reduce the strength of each vortex line while increasing their number: one may thus achieve a perfectly continuous velocity field, without local singularities.

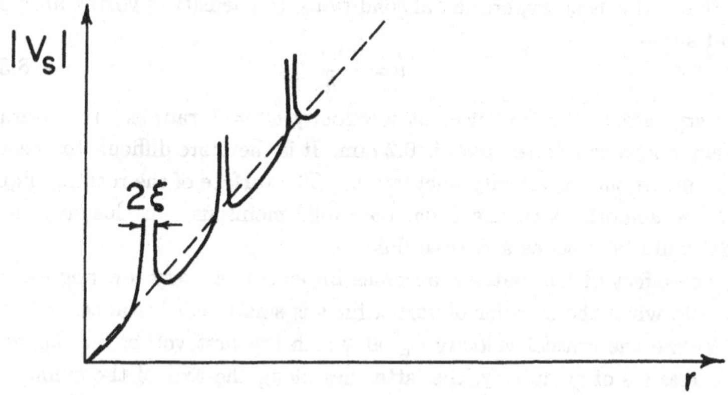

FIGURE 8.6. *Fluctuating velocity field in a rotating bucket.*

To the extent that we may neglect the local fluctuations of velocity, the kinetic energy of an array of vortex lines is simply $\frac{1}{2}I\omega_o^2$, where I is the moment of inertia of the liquid, equal to $MR^2/2$. To this we must add the kinetic energy due to the velocity fluctuations around each of the $\nu\pi R^2$ vortices in the bucket. We thus find

$$E \sim \frac{1}{2}I\omega_o^2 + \pi R^2 \nu \pi \frac{N}{m}\hbar^2 \log\frac{d}{\xi} \tag{8.31}$$

where d is the average distance between vortex lines. On making use of (8.30), we obtain

$$E \sim \frac{1}{2}I\omega_o^2 + \pi R^2 N\hbar\omega_o \log\frac{d}{\xi} \tag{8.32}$$

For reasonable values of R and ω_o, the second term of (8.32) is far smaller than the first: the fluctuations around each quantized vortex do not contribute appreciably to the kinetic energy.

In the same way, we may calculate the angular momentum $\mathbf{L}$ produced by the vortex lines. Within a small correction of order d, arising from the velocity fluctuations, $\mathbf{L}$ is simply equal to its classical value $I\omega_o$. Neglecting fluctuations, the minimum of $(E - \mathbf{L}\cdot\omega)$ is obtained for

$$\omega_o = \omega \tag{8.33}$$

On the average, the liquid is seen to rotate at the same angular velocity as the bucket, just as it would for a classical fluid. In order to maintain equilibrium, the whole pattern of vortex lines rotates along with the bucket, with the velocity ω.

Under the usual experimental conditions, the density of vortex lines per unit surface

$$\nu = \frac{m\omega}{\pi\hbar} \tag{8.34}$$

is very large. For instance, at a velocity $\omega = 1$ rad/sec, the average spacing between vortex lines is 0.2 mm. It is therefore difficult to observe the corresponding velocity fluctuations. The surface of the rotating liquid will be smooth, with the usual parabolic meniscus. In this respect, a superfluid behaves as a normal fluid.

The effect of fluctuations becomes important at very low angular velocities when the number of vortex lines is small. For instance, one may calculate the critical velocity ω_{c1} at which the first vortex line appears. For reasons of symmetry, the latter lies along the axis of the cylindrical bucket. Its energy is given by (8.4); its angular momentum $\mathbf{L}$ is readily calculated to be

$$\mathbf{L} = Nm \iint \mathbf{v} \times \mathbf{r} d\sigma = \pi R^2 N\hbar \tag{8.35}$$

Such an arrangement becomes energetically favorable when $(E - \mathbf{L}\cdot\omega) < 0$. The critical angular velocity is thus given by

$$\omega_{c1} = \frac{E}{L} = \frac{\hbar}{mR^2} \log \frac{R}{\xi} \tag{8.36}$$

We note that ω_{c1} is a decreasing function of the radius R of the vessel. Below the critical velocity, the stable configuration of the superfluid is that described in Chapter 4, in which the liquid is at rest, unaffected by the rotation of the bucket. The surface of the liquid should then be flat. Actually, such a behavior is difficult to observe experimentally, as ω_{c1} is extremely small. (For liquid helium, ω_{c1} is $\sim 10^{-3}$ rad/s for a bucket of radius $R = 1$ cm.)

In order to observe vortices in a rotating bucket, ω must be larger than the threshold value ω_{c1}. In principle, there should also exist an upper threshold, ω_{c2}, above which vortices are no longer formed in the bucket. This threshold is reached when the density of vortices is such that their cores touch one another. Since the core diameter is ξ, the coherence length, the corresponding vortex density is

$$\nu \sim \frac{1}{\xi^2}$$

According to (8.30), we thus have:

$$\omega_{c2} \sim \frac{h}{m\xi^2} \tag{8.37}$$

The frequency (8.37) is extremely high, of the order of the average roton frequency. The upper threshold, ω_{c2}, is thus well beyond the experimental range of angular velocities. Yet, the existence of such an upper threshold is interesting as a matter of principle: such a concept is essential in the description of hard superconductors.

8.5 Critical Velocity for Superfluid Flow[3]

We now return to the problem of the critical velocity for superfluid flow. The Landau criterion which we discussed in Chapter 5 predicts, for liquid helium, a critical velocity $v_c \sim 70$ m/s, which is independent of the size of the vessel. The observed critical velocities are far smaller, varying between 1 cm/s and 1 m/s: they furthermore depend on the size of the vessel, being maximum for very thin capillaries. Consequently, there is no doubt that the mechanism which led to the Landau criterion is not relevant in the actual breakdown of superfluid flow.

Feynman (1955) has proposed another mechanism by which superfluidity might be destroyed in a moving liquid, one which involves the formation of a large number of vortices, produced by the friction on the walls. However, although there appears to be little doubt that the critical velocity indeed depends on the appearance of vortices, the exact mechanism by which they act to damp superfluidity remains controversial. In what follows, we set forth a few qualitative remarks, without attempting to solve the problem.

We consider a cylindrical capillary, with a radius R, inside which the fluid flows at a uniform velocity v_s. When v_s exceeds a critical value v_c, we expect vortices to appear spontaneously in the liquid. We are then led to ask two questions:

1. How are the vortices formed in the liquid?
2. What happens to the liquid once the vortices are formed?

The mechanism by which vortices are created controls the critical velocity $\mathbf{v}_c$. What happens next determines the nature of "supercritical" flow.

Let us first consider the second question. We assume that a vortex ring is formed at the periphery of the capillary. Two things may happen:

1. The ring may shrink steadily, dissipating its energy into heat through mutual friction with the normal fluid (perhaps through

[3] For a review of critical velocity experiments and theory, see Vinen (1963).

excitation of vortex oscillations). When the ring becomes of an atomic size, it may eventually decay into one or several quasi-particles. Such a shrinking of vortex rings provides a mechanism by which the directed kinetic energy of the flow may be transformed into quasi-particle energy, i.e., into heat. If we force the fluid to flow at a velocity larger than $\mathbf{v}_c$, it will pass from the *superfluid* state to the *normal* state.

2. The vortex ring may become unstable, through the buildup of large amplitude vortex oscillations. The vortex line is then twisted in an essentially erratic way. One quickly reaches a situation where the vortices form a complicated tangle, characteristic of a *turbulent state.* Such a turbulence will certainly act to damp the superfluid flow, although the exact mechanism by which the damping occurs is far from clear. (For instance, a pressure head will be required to drive the fluid through the pipe.) It should, however, be recognized that in such a case, the supercritical regime does not correspond to a normal fluid flow, but rather to a *superfluid turbulence.* Put another way, $\mathbf{v}_c$ marks the onset of turbulence, not of viscosity in the usual sense.

Experimentally, it is found that the superfluid flow of liquid He II at velocities larger than $\mathbf{v}_c$ belongs to the second type, that of superfluid turbulence. That renders the corresponding theory very difficult (remember that even the simple case of normal turbulent fluids is not really understood). For that reason, we shall not attempt to describe the nature of the turbulent state. The interested reader is referred to the review article by Vinen (1961a) for further details on this problem both experimental and theoretical.

We now return to the first question: How are the vortices formed in the liquid? It is fairly clear that vorticity can be created only at the walls of the pipe, through the interaction of the moving fluid with the fixed walls. The vortex line then "peels off" the wall (a circumstance which certainly leads to a complicated tangle of crossing lines). The exact mechanism of vortex creation is still unknown; somehow, the fluctuations on an atomic scale give rise to a core of radius ξ, which subsequently moves away from the walls.

Even assuming that a vortex is formed, we still have to explain how it can move *into* the fluid. It is easily verified that a vortex line is *attracted* by the pipe walls, essentially because of the boundary condition that the normal component of the superfluid velocity vanish at the walls. (For a plane wall, the attraction may be described as the interaction with "an image vortex," symmetric with respect to the wall.) As a result of this attraction, the vortex must overcome a huge potential barrier in order to penetrate into the fluid. How it ever manages to pass that barrier remains another unsolved question. (Possible explanations might involve the wall irregularities or the repulsion between vortices.)

These few remarks clearly show that the creation of vortices in a moving superfluid represents a formidable theoretical problem. Challenging as it may be, it lies beyond the scope of this book. For that reason, we shall only consider a very crude model, which we expect to yield an order of magnitude estimate of the critical velocity.

We consider a circular vortex ring, centered on the pipe axis. Its radius r is necessarily $\leq R$, the pipe radius. We furthermore neglect the interaction of the vortex ring with the walls, and consider the ring as moving in an indefinite fluid (such an approximation is clearly not valid when $r \simeq R$). The *spontaneous* creation of the ring can occur only if it is energetically favorable, which means that the ring must have a negative energy in the "pipe" frame of reference. (We note that this condition, although necessary, is by no means sufficient.) In a frame of reference moving along with the fluid, the kinetic energy $E(\mathbf{r})$ associated with the vortex is roughly equal to its length, $2\pi r$, times the energy per unit length (8.4). We thus expect to find

$$E(\mathbf{r}) \sim 2\pi^2 r \frac{N\hbar^2}{m} \log \frac{r}{\xi} \tag{8.38}$$

where the upper cutoff of the log will be determined by the radius of the ring rather than by the vessel size. The energy in the pipe frame, $E'(\mathbf{r})$, is then given by the usual relation

$$E'(\mathbf{r}) = E(\mathbf{r}) + \mathbf{P} \cdot \mathbf{v}_s$$

where $\mathbf{P}$ is the momentum carried by the vortex. The calculation of $\mathbf{P}$ is a standard problem of classical fluid dynamics. One finds

$$\mathbf{P} = \pm 2\pi^2 N\hbar r^2 \tag{8.39}$$

(where again the $\pm$ sign characterizes the sense of rotation of the vortex). On making use of (8.38) and (8.39), we see that the condition for spontaneous vortex creation, $E'(\mathbf{r}) \leq 0$, reduces to

$$\frac{\hbar}{mr} \log \frac{r}{\xi} \pm v_s \leq 0 \tag{8.40}$$

Thus, only vortices corresponding to the minus sign can be formed; the vortex moves in a direction opposite to the main fluid flow.

The condition (8.40) is most easily satisfied when the vortex radius r is large. The latter, however, cannot exceed the pipe radius R. The first instability should thus appear when v_s reaches the "critical" value

$$v_c = \frac{\hbar}{mR} \log \frac{R}{\xi} \tag{8.41}$$

Such a derivation of the critical velocity is admittedly very crude, since it relies on a model which ignores many relevant features of the problem (such as, for instance, the image effect). Nevertheless, (8.41) is in fair agreement with the experimental data on helium; the observed dependence of v_c on the pipe radius R is of the predicted form. Actually, the success of (8.41) is mostly of a dimensional nature. After all, there are not so many parameters in the problem: any theory based on vortex formation is bound to give a critical velocity similar to (8.41). Therefore, the agreement of (8.41) with experiment does not substantiate our simple model (which is probably quite unphysical); it only proves that superfluid flow is limited by *some* vortex instability. The purpose of the above calculation was to provide an order of magnitude of v_c, not to describe what really happens.

CHAPTER 9

MICROSCOPIC THEORY: UNIFORM CONDENSATE[1]

Thus far we have been concerned primarily with macroscopic properties of superfluid Bose liquids in general, and of He II in particular. Where we have been concerned with microscopic properties, as in Chapter 3, we have discussed phenomenological theories of superfluid behavior. We now wish to turn our attention to microscopic theories of interacting Bose systems, that class of theories which begin with a given law of interaction between the particles, and derive the various properties of the system therefrom.

Microscopic theory does not, at present, provide a quantitative description of liquid He II. Indeed, there exists a satisfactory microscopic theory of an interacting boson system in only two limiting cases, those of

1. Weak repulsive interactions between the bosons.
2. A dilute gas with arbitrary repulsive interaction between the particles.

Nonetheless, a discussion of the microscopic theory of superfluid boson systems is of considerable interest. One sees, quite generally, how the macroscopic occupation of a single quantum state dictates the structure of the microscopic description of system behavior. Moreover, the above limiting cases provide explicit examples of "model" systems which behave in accord with the general principles we have set forth in the preceding chapters. In the present chapter we shall consider the equilibrium behavior of systems at $T = 0$, which, further, possess a uniform condensate;

[1]The treatment in this chapter follows closely that in Hugenholtz and Pines (1959) and Pines (1963).

in the following chapter, we consider a spatially-varying condensate. Our primary emphasis will be on the derivation of the elementary excitation spectrum and response functions for the model systems mentioned above. We shall not discuss microscopic theories of superfluid behavior at finite temperatures or under non-equilibrium conditions.

9.1 Basic Hamiltonian

The Hamiltonian for our system of interacting bosons may be written in the form:

$$H = \sum_{\mathbf{p}} \varepsilon_p^o a_{\mathbf{p}}^+ a_{\mathbf{p}} + \frac{1}{2} \sum_{\mathbf{pkq}} V_q a_{\mathbf{p+q}}^+ a_{\mathbf{k-q}}^+ a_{\mathbf{k}} a_{\mathbf{p}} \tag{9.1}$$

where $\varepsilon_p^o = p^2/2m$ is the free particle kinetic energy, and V_q is the Fourier-transform of the interaction potential between a pair of particles. The Hamiltonian (9.1) serves as a starting Hamiltonian in our consideration of microscopic theory; actually, we shall show that the existence of macroscopic occupation of a single quantum state leads to an effective Hamiltonian which looks rather different.

We assume that the particles are condensed in the state with zero momentum, and that the number of particles in that state is

$$N_o \gg 1 \tag{9.2}$$

As we have discussed in Chapter 2, the condensate is depleted as a result of the interaction between the particles; consequently N_o may be only a small fraction of N. Nonetheless, as long as that fraction is independent of the size of the system, the condition (9.2) is obviously well satisfied for a system of macroscopic size.

Because N_o is so large, one would expect that the addition or subtraction of a few particles in the zero-momentum state would not have any appreciable effect on the overall physical behavior of the system. Such a statement appears innocuous, as well as being nearly obvious. In fact, it has profound consequences for the development of a microscopic theory, as Bogoliubov first showed in 1947. To see how this comes about, we consider the behavior of the annihilation and creation operators for the zero-momentum state. When acting on the ground state wave-function, ψ_o, they yield:

$$a_o|\psi_o(N_o)\rangle = \sqrt{N_o}|\psi_o(N_o - 1)\rangle \tag{9.3}$$

$$a_o^+|\psi_o(N_o)\rangle = \sqrt{N_o+1}|\psi_o(N_o+1)\rangle \tag{9.4}$$

where $\psi_o(N_o)$ is the ground state wave-function with N_o particles in the condensate, $\psi_o(N_o-1)$ that with N_o-1 particles there, etc. Since $N_o \gg 1$, one may do two things:

1. Replace $\sqrt{N_o+1}$ by $\sqrt{N_o}$.
2. Assume that the difference between $\psi_o(N_o-1)$, $\psi_o(N_o)$, and $\psi_o(N_o+1)$ is of order $1/N$; that is, removing or adding a particle in the zero-momentum state *in no way* alters the physical properties of the system.

Under these circumstances, a_o and a_o^+ will commute with each other; since they already commute with all $a_{\mathbf{p}}$ and $a_{\mathbf{p}}^+$ for $\mathbf{p} \neq 0$, they may be replaced everywhere by the "c" number, $\sqrt{N_o}$. This is the famous Bogoliubov prescription; it amounts to treating the condensate as a static quantity and so neglecting any possible behavior it might have as a consequence of its coupling to an excited particle. This is clearly an excellent approximation: the *mass* of the condensate being of order N times the mass of any single particle, dynamic effects in particle-condensate interaction can scarcely be expected to play any role.

On applying the Bogoliubov prescription, one obtains the following Hamiltonian:

$$\begin{aligned} H = & \sum_{\mathbf{p}}{}' \left(\varepsilon_p^o + N_o V_o + N_o V_{\mathbf{p}}\right) a_{\mathbf{p}}^+ a_{\mathbf{p}} \\ & + \sum_{\mathbf{p}}{}' \frac{N_o V_p}{2} \left(a_{\mathbf{p}}^+ a_{-\mathbf{p}}^+ + a_{\mathbf{p}} a_{-\mathbf{p}}\right) + \frac{1}{2} N_o^2 V_o \\ & + \sum_{\mathbf{pq}}{}' \sqrt{N_o} \frac{V_q}{2} \left(a_{\mathbf{p+q}}^+ a_{\mathbf{q}} a_{\mathbf{p}} + a_{\mathbf{p+q}}^+ a_{-\mathbf{q}}^+ a_{\mathbf{p}}\right) \\ & + \sum_{\mathbf{pkq}}{}' \frac{V_q}{2} a_{\mathbf{p+q}}^+ a_{\mathbf{k-q}}^+ a_{\mathbf{k}} a_{\mathbf{p}} \end{aligned} \tag{9.5}$$

The primes are intended to remind the reader that nowhere do the creation and annihilation operators for the zero-momentum state appear. The zero-momentum state has disappeared from the problem, its only residue being the various factors of N_o which appear in (9.5). We see from the interaction terms in (9.5) that the number of bosons is not conserved.

There appear terms which create or destroy two bosons, scatter one into two, and vice-versa. Put another way, the particle number operator,

$$N'_{op} = \sum_{\mathbf{p}}' a^+_{\mathbf{p}} a_{\mathbf{p}} \tag{9.6}$$

which represents the number of particles which are *not* in the condensate, does not commute with the Hamiltonian, (9.5). We are taught in quantum mechanics that particle conservation is a Good Thing; to what extent, then, can one preserve it, while still making the quite physical approximation of treating the condensate as a fixed quantity?

Our original problem was to determine the ground state of the system of N interacting bosons, that is, the eigenstate of the Hamiltonian, (9.1), which has the ground state energy, E_o, subject to the condition that the number of particles is equal to N. In the modified problem, specified by the Hamiltonian (9.5), we must therefore impose the subsidiary condition

$$\langle N'_{op} \rangle = N'(N_o) = N - N_o \tag{9.7}$$

N_o then appears as a parameter, which is to be determined in such a way that the system energy is minimized [subject to the constraint (9.7)].

If N' commuted with H, one could satisfy the subsidiary condition (9.7), by imposing it on the unperturbed system wave-function. It would then automatically be satisfied for the true wave-function of the system. Such a possibility being not available, Hugenholtz and Pines (1959) have put forth an alternative method, based on a Lagrangian multiplier technique. They proposed that in place of (9.5) one consider the Hamiltonian,

$$H' = H(N_o) - \mu N'_{op} \tag{9.8}$$

while relaxing the subsidiary condition (9.7). The ground state wave-function appropriate to (9.8) is then a function of N_o and μ, $\psi'_o(N_o, \mu)$. The various expectation values of interest are likewise functions of both N_o and μ. The parameter μ, which turns out to be the chemical potential, is determined by the condition

$$N'(N_o, \mu) + N_o = N \tag{9.9}$$

where $N'(N_o, \mu)$ is the expectation value,

$$\langle \psi'_o(N_o, \mu) | N'_{op} | \psi'_o(N_o, \mu) \rangle = N'(N_o, \mu) \tag{9.10}$$

With such a choice of μ, $\psi_o'(N_o, \mu)$ is the ground state of (9.8) with the subsidiary condition (9.10). The remaining parameter, N_o, is determined by the condition that for fixed N,

$$\frac{dE_o}{dN_o} = 0 \tag{9.11}$$

A little algebra (cf. Hugenholtz and Pines) leads one to the following equivalent expression for μ,

$$\mu = \frac{\partial E_o'}{\partial N_o} = \frac{\partial E_o}{\partial N} \tag{9.12}$$

which shows that μ is indeed the chemical potential. The replacement of H by H', (9.8), is thus seen to be equivalent to replacing the particle kinetic energy, $p^2/2m$, by the value $p^2/2m - \mu$, that is, measuring the single particle energies from the chemical potential μ [which is given, in turn, by (9.12)].

The procedure we have described is needed whenever there is appreciable depletion of the zero-momentum state as a result of particle interaction. Such will certainly be the case for He II [we recall the estimate of a 92% depletion by Penrose and Onsager (1956)]. One need not, in fact, follow this procedure for the examples we treat in this section, since in both cases the depletion is quite small, and N_o may be simply replaced by N, as in the original work of Bogoliubov. We have gone into the above procedure in some detail because it illustrates clearly the extent to which particle-conservation must be abandoned in constructing a microscopic theory of superfluid behavior. We see that it is possible to maintain particle conservation on the average, whilst using a non-particle conserving Hamiltonian. Any departure from particle conservation is then unimportant, and must be ascribed to fluctuations in the condensate population.

The appearance of non-particle-conserving states is perhaps obvious in the case of a superfluid boson system. They are equally important for a fermion superfluid. We may say that superfluidity (boson *and* fermion) is characterized by macroscopic occupation of a single quantum state, and that such state occupation and lack of particle conservation go hand in hand. It is possible to construct, in principle, theories of superfluid behavior in which such states do not appear. In practice the resulting theories are clumsy and complicated and yield results in agreement with the non-conserving theory we consider here.

9.2 Weak Coupling Approximation

We consider now the solution of the Hamiltonian in a weak coupling approximation, due to Bogoliubov (1947). It is based on the following intuitive idea. In the ground state of the non-interacting system, all particles have zero momentum. Hence, when one takes into account interactions, those which are most important are those which involve the maximum number of zero-momentum particles. One can classify the interaction terms in $H(N_o)$ according to the powers of $\sqrt{N_o}$ they contain. On this basis, one is led to divide H', (9.8), into two parts,

$$H' = H_1 + H_2 \tag{9.13}$$

where

$$H_1 = \sum_{\mathbf{q}}{}' \tilde{\varepsilon}_{\mathbf{q}} a^+_{\mathbf{q}} a_{\mathbf{q}} + \sum_{\mathbf{q}}{}' N_o \frac{V_q}{2} \left(a^+_{\mathbf{q}} a^+_{-\mathbf{q}} + a_{\mathbf{q}} a_{-\mathbf{q}}\right) + \frac{1}{2} N_o^2 V_o \tag{9.14}$$

$$H_2 = N_o^{\frac{1}{2}} \sum_{\mathbf{pq}}{}' \frac{V_q}{2} \left(a^+_{\mathbf{p+q}} a_{\mathbf{q}} a_{\mathbf{p}} + a^+_{\mathbf{p+q}} a^+_{-\mathbf{q}} a_{\mathbf{p}}\right)$$

$$+ \sum_{\mathbf{pkq}}{}' \frac{V_q}{2} a^+_{\mathbf{p+q}} a^+_{\mathbf{k-q}} a_{\mathbf{k}} a_{\mathbf{p}} \tag{9.15}$$

and

$$\tilde{\varepsilon}_q = \varepsilon^o_q + N_o V_o + N_o V_q - \mu \tag{9.16}$$

H_1 is assumed to contain the most important parts of the particle interaction. H_2 gives rise to correction terms, which are assumed negligible in lowest order of approximation.

The Hamiltonian, H_1, may be diagonalized by means of a transformation to new variables, $\alpha_{\mathbf{p}}$ and $\alpha^+_{\mathbf{p}}$:

$$a_{\mathbf{p}} = u_p \alpha_{\mathbf{p}} - v_p \alpha^+_{-\mathbf{p}}$$
$$a^+_{\mathbf{p}} = u_p \alpha^+_{\mathbf{p}} - v_p \alpha_{-\mathbf{p}} \tag{9.17}$$

If u_p and v_p are real and satisfy the conditions

$$u_p^2 - v_p^2 = 1 \tag{9.18}$$

the transformation is readily seen to be canonical; the new creation and annihilation operators, $\alpha^+_{\mathbf{p}}$ and $\alpha_{\mathbf{p}}$, thus obey the usual boson commutation rules:

$$\left[\alpha_{\mathbf{p}}, \alpha^+_{\mathbf{p}'}\right] = \delta_{\mathbf{p},\mathbf{p}'}$$

$$\left[\alpha_{\mathbf{p}}, \alpha_{\mathbf{p}'}\right] = \left[\alpha^+_{\mathbf{p}}, \alpha^+_{\mathbf{p}'}\right] = 0 \tag{9.19}$$

The reader may readily verify that in the new representation the off-diagonal terms in H are eliminated if one takes:

$$u_q^2 = \frac{1}{2}\left[1 + \frac{\tilde{\varepsilon}_q}{\omega_q}\right] \tag{9.20}$$

$$v_q^2 = \frac{1}{2}\left[\frac{\tilde{\varepsilon}_q}{\omega_q} - 1\right] \tag{9.21}$$

where

$$\omega_q^2 = \tilde{\varepsilon}_q^2 - N_o^2 V_q^2 \tag{9.22}$$

H_1 then takes the following form:

$$H_1 = {\sum_{\mathbf{q}}}' \left\{\omega_q\, \alpha^+_{\mathbf{q}}\alpha_{\mathbf{q}} + \frac{1}{2}\left(\omega_q - \tilde{\varepsilon}_q\right)\right\} + \frac{1}{2}N_o^2 V_o \tag{9.23}$$

The Bogoliubov transformation leads to a new set of *independent* quasi-particles, which are characterized by the creation operators, $\alpha^+_{\mathbf{q}}$, and energy ω_q. The ground state of the system is defined as the "vacuum" of such quasi-particles, and is thus specified by the relation

$$\alpha_{\mathbf{q}}|0\rangle = 0 \tag{9.24}$$

In order to completely determine ω_q, we must calculate the chemical potential μ. This is easily done to lowest order in V_q: on making use of (9.12) and noting that the leading term in the ground state energy is $1/2\,(N_o^2 V_o)$, we find

$$\mu = N_o V_o \tag{9.25}$$

Hence

$$\tilde{\varepsilon}_q = \frac{q^2}{2m} + N_o V_q \tag{9.26}$$

while

$$\omega_q = \left(\frac{q^2 N_o V_q}{m} + \frac{q^4}{4m^2}\right)^{1/2} \tag{9.27}$$

Since V_q approaches a constant, V_o, for small q, the new quasi-particles are seen to be phonons in the long wave-length limit, with a velocity

$$s_B = \left(\frac{N_o V_o}{m}\right)^{1/2} \tag{9.28}$$

At short wave-lengths the quasi-particles behave like almost free particles, with an energy

$$\varepsilon_{\text{HF}} = \frac{q^2}{2m} + N_o V_q = \tilde{\varepsilon}_q \tag{9.29}$$

This latter energy is just the energy a quasi-particle would have in the Hartree-Fock approximation, since, in this approximation the second term of H_1 may be neglected altogether.

We see that the dramatic new features of the boson system appear at long wave-lengths, specifically for

$$\lambda \gg \lambda_c = \left(\frac{1}{2ms_B}\right) \tag{9.30}$$

λ_c is the characteristic length h for the transition from phonon-like behavior to single particle-like behavior. The fact that in a consistent microscopic theory the long wave-length quasi-particles would be phonons was certainly to be expected on the basis of our derivation in Chapter 2. The nature of the quasi-particles in this limit could not, however, have been anticipated. On inverting the transformation, (9.17), we find

$$\alpha_{\mathbf{q}}^{+} = u_q \, a_{\mathbf{q}}^{+} + v_q \, a_{-\mathbf{q}} \tag{9.31}$$

while

$$\lim_{q \to 0} u_q = v_q = \sqrt{\frac{ms_B}{2q}} \tag{9.32}$$

Hence a long wave-length quasi-particle of momentum q is nearly in equal parts a "bare" particle of momentum q and a bare "hole" of momentum $-q$.

Such behavior contrasts sharply with that of a normal fermion system. There we saw that a quasi-particle consists in configurations involving a bare particle, two bare particles and one hole, etc., all of which have the same particle number. The present situation reflects clearly the lack of particle conservation. The interaction term in H_1 which is responsible is that which creates or destroys two bare particles; by means of it, a particle of momentum q may be transformed, in part, into a hole of momentum $-q$. At small q this term dominates the behavior of the system particles to such an extent that the resulting quasi-particle is equally shared between a bare particle and a bare hole. This aspect of the Bogoliubov theory is common to all existing microscopic theories. It shows clearly how the condensate acts as a reservoir (or sink) of particles.

The depletion of the condensate as a result of particle interaction is readily calculated. One finds:

$$N - N_o = \langle 0| \sum_{\mathbf{p}}{}' a_{\mathbf{p}}^{+} a_{\mathbf{p}} |0\rangle = \sum_{\mathbf{p}} v_p^2 = \frac{1}{2}\sum_{\mathbf{p}} \left\{ \frac{\varepsilon_p + N_o V_p}{\omega_p} - 1 \right\} \tag{9.33}$$

It is seen to be of order V_p, and hence is negligible to the extent that the interaction is weak. We may, therefore, replace N_o by N everywhere it appears in the foregoing (apart from the lowest-order Hartree term in E_o).

It is left as an exercise to the reader to show that the dynamic form factor is given by:

$$s_B(\mathbf{q}, \omega) = \frac{N_o q^2}{2m\omega_q} \delta(\omega - \omega_q) \simeq \frac{N q^2}{2m\omega_q} \delta(\omega - \omega_q) \tag{9.34}$$

One verifies readily that $s_B(\mathbf{q}, \omega)$ satisfies the sum rules, (2.20) and (2.21) [in computing the right-hand side of the latter, one works only to lowest order in μ; that is μ is given by (9.25)]. The density-density response function is:

$$\chi_B(\mathbf{q}, \omega) = \frac{N q^2/m}{\omega^2 - \omega_q^2} \tag{9.35}$$

We see that for all values of $\mathbf{q}$, one has only single quasi-particle excitations contributing to both $S(\mathbf{q}, \omega)$ and $\chi(\mathbf{q}, \omega)$, a feature which is special to the Bogoliubov approximation. There are no contributions due to multi-particle excitations from the condensate, because such arise only via H_2, which we have neglected. The correlation length defined via (9.30) is therefore not obviously the same as either of the coherence lengths we have introduced in Chapter 4; in fact, closer investigation shows that it is identical to both for the case of a pure system.

We have asserted that the Bogoliubov approximation of keeping only H_1 is a weak coupling approximation. It is not that in the normal sense in which, say, one keeps only the second-order term in the particle interaction. Rather it is the lowest-order *consistent* weak-coupling approximation and is, in fact, equivalent to the random phase approximation. We now develop this thesis in some detail.

From the standpoint of perturbation theory [Brueckner and Sawada (1957)], the lowest order calculation is the Hartree-Fock approximation. In this, the ground state wave function is

$$\psi_{\mathrm{HF}} = (a_o^{+})^{N_o} |0\rangle \tag{9.36}$$

where $|0\rangle$ is a vacuum state in which there are no particles present. The ground state energy and quasi-particle energy are easily seen to be

$$E_{\mathrm{HF}} = \frac{1}{2}N_o^2 V_o = \frac{1}{2}N^2 V_o \tag{9.37}$$

$$\varepsilon_q^{\mathrm{HF}} = \frac{q^2}{2m} + N_o V_q \tag{9.38}$$

To calculate the next higher order terms in the perturbation-theoretic expansion of H, one must decide which of the various interaction terms in the Hamiltonian, (9.13), gives a non-vanishing result when acting on the state, (9.36). There is only one:

$$\sum_{\mathbf{q}}{}' \left(\frac{V_q}{2}\right) a_{\mathbf{q}}^+ a_{-\mathbf{q}}^+ \tag{9.39}$$

It gives rise to a shift in the ground state energy which is:

$$E^{(2)} = \sum_q{}' \frac{N_o V_q^2 m}{q^2} \tag{9.40}$$

and which may be represented by the diagram shown in Fig. 9.1a. There one sees the excitation of a pair of particles, via (9.39), and their subsequent de-excitation due to its complex conjugate.

All terms in the perturbation series expansion of the ground state energy of H' can easily be represented with the aid of such diagrams. A third-order perturbation-theory term, which arises in H_1, is shown in Fig. 9.1b. There one sees pair excitation, followed by forward scattering of one of the particles, then by pair annihilation. The forward scattering matrix element is

$$N_o V_o + N_o V_q - \mu = N_o V_q \tag{9.41}$$

and the contribution made by this term to the ground state energy is

$$E_a^{(3)} = -\sum_q{}' \frac{(N_o V_q)^3}{(q^2/m)^2} \tag{9.42}$$

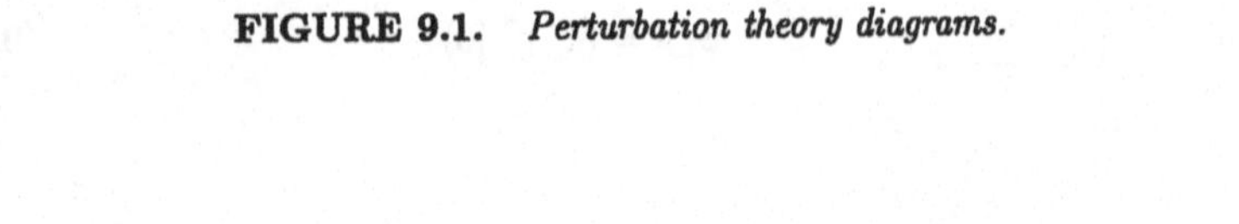

FIGURE 9.1. *Perturbation theory diagrams.*

It is divergent for small momentum transfers (since V_q is then a constant), so that a theory which stops at this point is clearly inconsistent.

Both contributions we have thus far considered come from H_1; there is a further third-order term which comes from H_2, and which is not divergent. It is shown in Fig. 9.1c and is given by

$$E_b^{(3)} = \sum_{\mathbf{kq}}{}' \frac{N_o^2 V_q V_k V_{\mathbf{q-k}}}{(q^2/m)(k^2/m)} \tag{9.43}$$

It differs from (9.40) in that only two powers of N_o appear, and particle interactions which involve momentum other than **q** are taken into account. Study of the higher-order terms in the series shows that, in each order, those terms in which only the *single* momentum transfer **q** appears possess the highest powers of N_o, and are the most divergent. These are just the terms which arise in H_1; they correspond to diagrams of the type shown in Fig. 9.1d. If one sums the *entire* series of terms coming from H_1, one obtains for the ground state energy,

$$E_o = \sum_{\mathbf{q}}{}' \left\{ \frac{\omega_q}{2} - \frac{\varepsilon_q^o + N_o V_q}{2} \right\} + \frac{1}{2} N_o^2 V_o \tag{9.44}$$

which is identical to the lowest energy state of the quasi-particle Hamiltonian, (9.14).

The Bogoliubov transformation is thus seen to be an easy way to sum an entire set of terms in the perturbation series expansion of H, those which involve repetitions of only a single momentum transfer, **q**. The alert reader will now remark that the Bogoliubov approximation, in which H_2 is neglected, must be equivalent to the random phase approximation, since in the RPA one takes into account all effects associated with a single momentum transfer **q**. Physically it corresponds to including, to lowest order, the effects of the time-dependent self-consistent field produced by all the other particles on the motion of a given particle.

The connection between the Bogoliubov approximation and the RPA may be made explicit with the aid of one of the many different formulations of the RPA presented in Chapter 5, Vol. I. There we saw that if in the Hartree approximation the density-density response function is $\chi_o(\mathbf{q},\omega)$, then in the RPA it will necessarily be:

$$\chi_{\text{RPA}}(\mathbf{q},\omega) = \frac{\chi_o(\mathbf{q},\omega)}{1 - V_q \chi_o(\mathbf{q},\omega)} \tag{9.45}$$

For the Bose system, one sees from the definition of $\chi(\mathbf{q},\omega)$ that the lowest order (free boson or Hartree) approximation is simply:

$$\chi_o(\mathbf{q},\omega) = \frac{Nq^2/m}{\omega^2 - q^4/4m^2} \tag{9.46}$$

On substituting (9.46) into (9.45), one finds at once the earlier result, (9.35). It is left as a problem to the reader to show that one may pass directly from (9.45) to the result obtained from (9.23) and (9.24) for the ground-state energy.

We return now to the question of the range of validity of the Bogoliubov approximation (the RPA). To decide this, one needs to consider the relative importance of the terms in the particle interaction which have been neglected, namely, those given by H_2. Systematic investigations of higher order terms are best carried out by field-theoretic methods, which lie beyond the purview of this book. In the present case, one finds [Beliaev (1958)] that the RPA is a *weak-coupling* theory, which for a short-range interaction is valid in the limit of $V_p \rightarrow 0$. Let us emphasize that the RPA is the lowest-order (in V_p) *consistent* theory of an interacting boson system, in much the same way that it is for the electron system. In the present context, by consistent we mean a theory which yields results in accord with the exact long wave-length behavior we have established earlier. The Hartree-Fock approximation for example, is not a consistent approximation, since one finds within it an energy gap (N_oV_o) for a long wave-length quasi-particle excitation, whereas we have shown that the long wave-length quasi-particles must be phonons. By contrast, in the RPA, the low momentum quasi-particles are phonons, and the dynamic form factor satisfies both the f-sum rule and the compressibility sum rule, (2.20) and (2.21), to lowest order in V_p.

To summarize: the Bogoliubov approximation, although valid only in a weak-coupling limit, offers the essential clue to a microscopic theory of superfluidity. One sees clearly how the structure of the theory is dictated by the existence of the condensate: one finds that the long wave-length excitations are phonons, with a velocity equal to the macroscopic sound velocity computed to lowest order of approximation.

Throughout this section we have tacitly assumed that $V_q > 0$, that is, that the particle interaction is repulsive. If this is not the case, and if

$$|NV_q| > \frac{q^2}{4m}$$

one sees readily that the corresponding quasi-particle excitations are unstable. For example, if $V_o < 0$, the corresponding long wave-length phonons are unstable. In this limit, the physical origin of the instability is clear; under such circumstances the compressibility is no longer positive definite, so that the system will simply collapse. We note that quite generally an attractive interaction can easily create problems in the treatment of interacting boson systems, essentially because, in contrast to fermions, there is no large positive kinetic energy to stabilize the system behavior. We return to this general question later on in the chapter.

9.3 Dilute Bose Gas

The dilute Bose gas provides a well-defined *model* for the behavior of a superfluid boson system. In this sense it resembles the high density electron gas and the dilute neutral fermion gas. All three systems are "model" systems in the sense that, in suitable limit, one can solve as accurately as one likes for the various quantities of physical interest. An accurate solution is possible for the high density electron gas because in that limit the potential energy is small compared to the kinetic energy. For dilute systems, accuracy of approximation depends on the ratio

$$\frac{f_o}{r_o} \ll 1 \tag{9.47}$$

Here f_o is the s-wave scattering length for two free particles, while r_o is the average inter-particle spacing.

In order to understand the physical basis for the approximation, let us consider a collision between two particles. If the system is sufficiently dilute, such collisions are not frequent. When they occur, the chances that a third particle be involved in the process are very small. Thus, to a high degree of approximation, we can consider that the only effect of the particle interactions is to produce *binary collisions*. During one such collision, the pair of colliding particles is not even aware of the fact that it is part of a larger many-particle system.

In a weakly interacting gas, we might describe binary collisions between a pair of particles within the Born approximation. If, however, the interaction potential is strong, the Born approximation is inapplicable. In such a case, one must treat the scattering of a pair of particles *exactly* by taking due account of *multiple scattering* effects. Because the system is dilute, multiple scattering of one particle against the other may be expected to be identical to that for two particles in vacuo. For two low

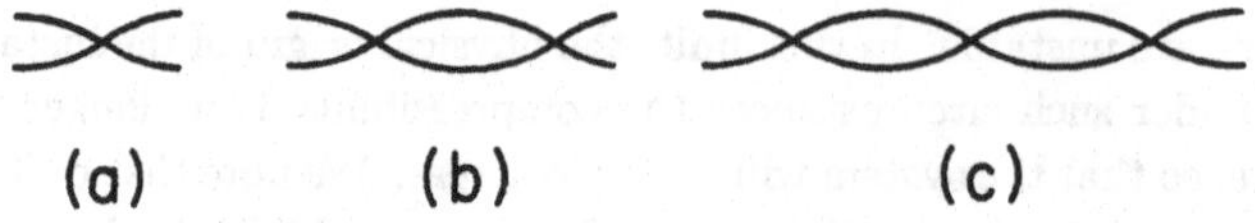

FIGURE 9.2. *Multiple scattering diagrams.*

energy particles, the effect of such multiple scattering is completely specified by the s-wave scattering length f_o. Qualitatively, the condition that a low density approximation be valid is that f_o be small compared to the only other length in the problem, the interparticle spacing r_o.

Two steps are required to obtain the lowest approximation to the ground state energy and excitation spectrum of a dilute Bose gas. First, one must account for the multiple scattering effect we have described above. What is needed may be visualized quite simply from the diagrams shown in Fig. 9.2. There we see two particles which scatter against one another once, twice, and three times. The effective interaction between a pair of particles, which takes into account *all* such simple multiple scattering events, is known as the two-particle scattering matrix, or "t" matrix. It is determined by the diagrammatic equation, shown in Fig. 9.3. It is not difficult to solve this equation [see, for example, Pines (1962), p. 75]; one finds that to lowest order in the density of the particles, the effective interaction between a pair of particles is a contact interaction:

$$t(\mathbf{r}-\mathbf{r}') = \frac{f_o}{m}\delta(\mathbf{r}-\mathbf{r}') \tag{9.48}$$

The prescription, then, for taking into account the repeated scattering between a single pair of particles, is to replace V_q wherever it may appear by f_o/m.

The second step of the calculation is simply to solve the *resulting* problem within the RPA; one is forced to do this, in a sense, since low-order

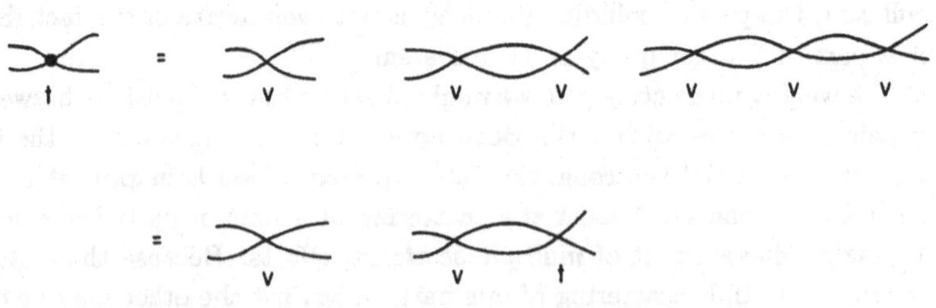

FIGURE 9.3. *Diagrammatic equation for the "t" matrix.*

perturbation theory using the interaction, (9.48), is no more successful than that based on the bare interaction, V_q. The Bogoliubov transformation is required to take into account the fact that one is dealing with a many-boson problem, the main features of which are set by the macroscopic occupation of the zero-momentum state.

It is, of course, trivial to substitute f_o/m for V_q everywhere in the results which have been obtained for the weakly coupled Bose system. There is, however, one caution which must be observed: in so doing one must be careful not to allow for the same physical effect twice. Here diagrams are useful since they show clearly what sorts of scatterings have been considered, what not. In the above two steps, we sum simultaneously the RPA and the multiple scattering (or ladder) diagrams for the interacting boson system. Put another way, we sum *all* RPA diagrams, *each* interaction vortex being replaced by a full series of ladder diagrams. A little study shows that one gets into no difficulties as far as the excited states of the system are concerned. There is, however, a duplication of diagrams (e.g., physical effects) in the calculation of the ground state energy. This may be seen very easily by inspection of the first two terms in the RPA summation, corresponding respectively to the Hartree approximation and to second order perturbation theory. The corresponding diagrams, using the effective interaction (9.48), are shown on Fig. 9.4. We see that the second order term completely duplicates terms which are already included in the Hartree term. In order to avoid such duplication in the calculation of the ground state energy, we must therefore subtract the second order term

$$E^{(2)} = -\sum_{\mathbf{q}}{}' \frac{N_o f_o^2}{mq^2} \tag{9.49}$$

from the straightforward RPA result. The ground state energy is, therefore,

$$E_o = \frac{1}{2}\sum_{\mathbf{q}}{}' \left\{ \left(\frac{N_o q^2 f_o}{m^2} + \frac{q^4}{4m^2} \right)^{1/2} - \frac{q^2}{2m} - \frac{N_o f_o}{m} \right\} + \frac{1}{2}\frac{N_o^2 f_o}{m} + \sum_{\mathbf{q}}{}' \frac{N_o f_o^2}{mq^2} \tag{9.50}$$

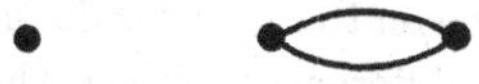

FIGURE 9.4. *Diagram duplication.*

Some properties of the dilute Bose gas, as calculated to lowest order in the scattering amplitude f_o, are as follows:

Excitation spectrum:

$$\omega_q = \left(\frac{N_o q^2 f_o}{m^2} + \frac{q^4}{4m^2}\right)^{1/2} \tag{9.51}$$

Sound velocity:

$$s = \frac{(N_o f_o)^{1/2}}{m} \tag{9.52}$$

Coherence length:

$$\lambda_c = \frac{1}{2} ms = \frac{1}{2}(N_o f_o)^{1/2} \tag{9.53}$$

Depletion of the ground state:

$$N = N_o \left[\frac{1 + (N f_o^3)^{1/2}}{3\pi^2}\right] \tag{9.54}$$

Ground state energy:

$$E_o = \frac{1}{2}\frac{N^2 f_o}{m}\left[1 + \frac{16}{15\pi^2}(N f_o^3)^{1/2} + \cdots\right] \tag{9.55}$$

Density-density correlation function:

$$\chi(\mathbf{q}, \omega) = \frac{Nq^2/m}{\omega^2 - \omega_q^2} \tag{9.56}$$

The above results were first obtained by Lee, Huang, and Yang (1957), who used, in fact, a somewhat different method than that we have sketched here. Their calculations were based on the Fermi pseudo-potential method, in which as a first approximation the effective potential is taken to be

$$V_{\text{eff}} = \frac{f_o}{m}\frac{\partial}{\partial r}[r\delta(r)] = \frac{f_o}{m}\delta(r) + r\frac{f_o}{m}\frac{\partial}{\partial r}\delta(r) \tag{9.57}$$

The second term in (9.57) is of just such a character that it leads to the addition of the term, $\sum_q' N_o f_o^2/mq^2$, when one carries out the Bogoliubov transformation. In the method we have presented, the effective potential is simpler, but one must be on the alert for diagram duplication, the latter effect accounting for the difference between the two effective potentials, (9.48) and (9.57).

There is one somewhat surprising feature of the results, (9.51) to (9.55): it is clear from the latter two equations that the basic expansion parameter is not f_o/r_o, but rather

$$\left(Nf_o^3\right)^{1/2} \tag{9.58}$$

The latter parameter is the ratio of f_o to the coherence length, $(N_o f_o)^{-1/2}$; it is this ratio which is important because one must, at the outset, include RPA correlations in order to get a consistent theory for the interacting boson system.

We note that in the above lowest order approximation to the dilute Bose gas, there are again no contributions to the density-density (and other) correlation functions from multiparticle excitations out of the condensate. Thus $\chi(\mathbf{q},\omega)$ takes the form, (9.56), for all values of $\mathbf{q}$, while $\chi_\perp(\mathbf{q},o)$ vanishes for all values of $\mathbf{q}$, not simply the small ones. This will no longer be the case if one calculates the various system properties to the next order in $\left(Nf_o^3\right)^{1/2}$. Higher order calculations of the excitation spectrum and the ground state energy have been carried out by a number of authors [Beliaev (1958), Lee and Yang (1958), Hugenholtz and Pines (1959), Sawada (1959), and Wu (1959)]. The results are:

$$E_o = \frac{N^2 f_o}{2m}\left[1 + \frac{16}{15\pi^2}\left(Nf_o^3\right)^{1/2} + \frac{Nf_o^3}{8\pi^2}\left(\frac{4}{3} - \frac{3}{\pi}\right)\ln Nf_o^3 + \cdots\right] \tag{9.59}$$

and

$$\omega_q = \frac{q}{m}\left(Nf_o\right)^{1/2}\left\{1 + \frac{1}{\pi^2}\left(Nf_o^3\right)^{1/2} + \cdots - i\frac{3}{640\pi}\left(Nf_o^3\right)^{1/2}\frac{q^4}{\left(Nf_o\right)^2}\right\} \tag{9.60}$$

the latter result applying only in the low momentum region $\left(q \ll \sqrt{Nf_o}\right)$. The appearance of the logarithmic term in (9.59) shows that there does not exist a simple power series expansion in $\left(Nf_o^3\right)^{1/2}$ for the various properties of the dilute Bose gas. The imaginary term in the phonon energy arises because for this system it is energetically possible for a long wave-length phonon to decay into two phonons. The decay goes as q^5 because of the appearance of numerous coherence factors; it is obviously not very important at long wave-lengths.

We note, again, that the microscopic sound velocity agrees with that calculated macroscopically from the ground state energy, provided one keeps in E_o terms of the right order. It is always characteristic of a perturbation series expansion that if one calculates (microscopically) the excitation spectrum to a given order, one automatically obtains the ground state energy to one higher order. For example, the Bogoliubov approximation yields the lowest-order sound wave velocity, whilst giving at the same time the ground state energy (and the macroscopic sound velocity) to one higher order in $(Nf_o^3)^{1/2}$.

Detailed microscopic calculations of the other properties of the dilute Bose gas to next order in $(Nf_o^3)^{1/2}$ are feasible, but clearly lengthy, and have not as yet been carried out. Nonetheless, one may be confident, from the perturbation-theoretic studies cited above, that the lowest-order results we have quoted are correct as long as $(Nf_o^3)^{1/2} \ll 1$. Thus the dilute gas at $T = 0$ offers an example of a model system which has been shown to display superfluid behavior in complete accord with the Bose liquid theory developed in Chapters 2 and 4. In principle, one might use it to verify (in a particular example) the predictions of Chapter 5, although such calculations have not been carried out in detail.

The properties of the dilute Bose gas at finite temperatures have not yet been worked out in the same detail as the corresponding properties at $T = 0$. We shall therefore not discuss them here.

9.4 Application to He II

It is natural to inquire whether there exists an approximate microscopic theory of liquid He II. Considerable progress has been made in variational calculations of the ground state energy, compressibility, and pair distribution function. On the other hand, very little progress has been made on calculations which might be regarded as a direct extension to He II of those described in the preceding sections. Thus no one has yet been able to show that a given class of diagrams in the perturbation series are particularly relevant to the behavior of He II, and that if one takes these into account one arrives at an excitation spectrum which agrees with experiment. We comment briefly on some of the difficulties which arise when one attempts to carry out such a program.

First, it is clear that the dilute Bose gas is not especially relevant to He II. For example, the coherence length of the former is $(Nf_o)^{-1/2}$; it is large compared to the interparticle spacing, whilst for He II, the two lengths are roughly comparable. If one takes for f_o the scattering length appropriate to the repulsive part of the interaction between He atoms, one sees that $Nf_o^3 \sim 1$, so that the series expansion in $(Nf_o^3)^{1/2}$ breaks down completely. Finally, there is an attractive part of the interaction between He atoms, which plays an unknown role. We may speculate that it is not so strong as to lead to a negative value of f_0, since the compressibility of He II is positive.

Nonetheless, microscopic field theoretic methods have proved useful in the general study of Bose liquids. With their aid, it has been possible to show that in the long wave-length limit, the quasi-particle excitations cannot possess an energy gap [Hugenholtz and Pines (1959)]. Moreover, in this same limit, the validity of the expression (2.28), for $S(\mathbf{q}, \omega)$, has been established by a detailed microscopic calculation, as has, likewise, the result, (4.10), for $\chi_\perp(\mathbf{q}, 0)$ [Gavoret and Nozières (1963)]. Finally, as we have mentioned, corrections to the simple Landau expression for ρ_n have been obtained using microscopic methods [Balian and de Dominicis (1971)].

CHAPTER 10

MICROSCOPIC THEORY: NON-UNIFORM CONDENSATE

We consider in this chapter the extension of the microscopic theory of Chapter 9 to the case of a condensate structure which varies in space and time. The theory displays clearly the microscopic origin of the macroscopic theory of condensate motion presented in Chapter 5. It provides as well a description of states in which the condensate motion varies so rapidly that the methods of Chapter 5 are inapplicable.

10.1 Condensate Wave Function

The first, obvious problem is to find a suitable *definition* of the condensate structure. For that purpose, we consider the quantity

$$\rho(\mathbf{r}, \mathbf{r}') = \langle \varphi | \psi^*(\mathbf{r}) \psi(\mathbf{r}') | \varphi \rangle \tag{10.1}$$

where ψ^* and ψ are the creation and destruction operators for particles at points $\mathbf{r}$ and $\mathbf{r}'$; the average is taken in the state $|\varphi\rangle$ which we want to describe. $\rho(\mathbf{r}, \mathbf{r}')$ is called the *density matrix* of the system; it serves to characterize the correlations between particles located at $\mathbf{r}$ and $\mathbf{r}'$. By introducing a complete kit of eigenstates $|\varphi_n\rangle$, we may express (10.1) in an expanded form which is useful for studying the structure of $\rho(\mathbf{r}, \mathbf{r}')$:

$$\rho(\mathbf{r}, \mathbf{r}') = \sum_n \langle \varphi | \psi^*(\mathbf{r}) | \varphi_n \rangle \langle \varphi_n | \psi(\mathbf{r}') | \varphi \rangle \tag{10.2}$$

We consider first the density matrix $\rho_o(\mathbf{r}, \mathbf{r}')$ in the ground state $|\varphi_o\rangle$. The corresponding excitation spectrum has been discussed in detail in

Chapter 9; because of translational invariance, each excited state $|\varphi_n\rangle$ has a well-defined momentum $\mathbf{k}$, which is a constant of the motion. We denote such a state by the symbol $|\varphi_{n\mathbf{k}}\rangle$. Let us expand $\psi(\mathbf{r})$ in the usual way

$$\psi(\mathbf{r}) = \sum_{\mathbf{k}} a_{\mathbf{k}} e^{i\mathbf{k}\cdot\mathbf{r}} \tag{10.3}$$

On making use of translational invariance, we may write (10.2) in the form

$$\rho_o(\mathbf{r}, \mathbf{r}') = \sum_{\mathbf{k}} \langle \varphi_o | a_{\mathbf{k}}^* | \varphi_{n\mathbf{k}} \rangle \langle \varphi_{n\mathbf{k}} | a_{\mathbf{k}} | \varphi_o \rangle \, e^{i\cdot\mathbf{k}\cdot(\mathbf{r}'-\mathbf{r})} \tag{10.4}$$

We see that $\rho_o(\mathbf{r}, \mathbf{r}')$ contains essentially two types of terms, depending on which intermediate state $|\varphi_{n\mathbf{k}}\rangle$ is under consideration:

1. When $\mathbf{k} = 0$, the state $|\varphi_{n\mathbf{k}}\rangle$ is the ground state of an $(N-1)$ particle system, obtained by pulling one particle out of the condensed state, $\mathbf{k} = 0$. The corresponding matrix element is

$$\langle \varphi_o(N-1) | a_o | \varphi_o(N) \rangle = \sqrt{n_o} \tag{10.5}$$

   where n_o is the number of condensed particles. The contribution of that intermediate state to $\rho_o(\mathbf{r}, \mathbf{r}')$ is equal to n_o; it is of *macroscopic* size and must be treated separately.

2. All the other intermediate states $|\varphi_{n\mathbf{k}}\rangle$ give rise to matrix elements

$$\langle \varphi_{n\mathbf{k}} | a_{\mathbf{k}} | \varphi_o \rangle \sim 1 \tag{10.6}$$

   The number of such states is very large. Thus, although each individual state provides a term of order 1, the whole set of excited states $|\varphi_{n\mathbf{k}}\rangle$ yields a contribution to $\rho_o(\mathbf{r}, \mathbf{r}')$ which is of order N, and which thus can compete with the contribution of the single state $|\varphi_o(N-1)\rangle$.

A simple alternative expression for $\rho_o(\mathbf{r}, \mathbf{r}')$ is obtained by inserting the expansion, (10.3), directly into (10.1). One finds

$$\rho_o(\mathbf{r}, \mathbf{r}') = \sum_{\mathbf{k}} n_{\mathbf{k}} e^{i\mathbf{k}\cdot(\mathbf{r}-\mathbf{r}')} \tag{10.7}$$

which may be written as

$$\rho_o(\mathbf{r}, \mathbf{r}') = n_o + \bar{\rho}_o(\mathbf{r}, \mathbf{r}') \tag{10.8a}$$

where

$$\bar{\rho}_o(\mathbf{r}, \mathbf{r}') = \sum_{\mathbf{k}}{}' n_{\mathbf{k}} e^{i\mathbf{k}\cdot(\mathbf{r}-\mathbf{r}')} \tag{10.8b}$$

(as usual, the prime on the summation indicates that we are excluding the state $\mathbf{k} = 0$). The first term, n_o, on the right-hand side of (10.8a), is clearly associated with the presence of a condensate. The other term, $\bar{\rho}_o(\mathbf{r}, \mathbf{r}')$, arises from all the *excited* states $|\varphi_{n\mathbf{k}}\rangle$.

The function $\rho_o(\mathbf{r}, \mathbf{r}')$ is seen to depend only on the difference $(\mathbf{r} - \mathbf{r}')$: such simple behavior arises as a consequence of translational invariance and is a peculiar feature of the ground state. When $\mathbf{r} - \mathbf{r}' = 0$, $\bar{\rho}_o$ reduces to

$$\bar{\rho}_o(0) = \sum_{\mathbf{k}}{}' n_{\mathbf{k}} = N - n_o \; ; \tag{10.9}$$

it measures the depletion of the condensate arising from particle interactions. In the physical case of reasonably strong interactions, $\bar{\rho}_o(0)$ is comparable to n_o (in liquid ^{4}He, it may be larger by an order of magnitude). In the opposite limit, $(\mathbf{r} - \mathbf{r}') \to \infty$, the function $\bar{\rho}_o(\mathbf{r} - \mathbf{r}')$ vanishes as a consequence of *destructive interference* between the various phase factors present in the definition (10.8). In contrast, the first term of the density matrix, n_o, remains constant when $(\mathbf{r} - \mathbf{r}') \to \infty$. Such different behavior arises because in one case we have a *continuous* spectrum of intermediate states, with varying $\mathbf{k}$, while in the other we have a *discrete* intermediate state with $\mathbf{k} = 0$.

We summarize the essential features displayed by the preceding discussion. In a superfluid system, the density matrix in the ground state, $\rho_o(\mathbf{r}, \mathbf{r}')$ may be written in the form (10.8). The first term, n_o, is of *infinite range* in $(\mathbf{r} - \mathbf{r}')$; it arises directly as a consequence of the *macroscopic occupation of a single quantum state, the condensate.* The second term, $\bar{\rho}_o(\mathbf{r} - \mathbf{r}')$, has a *finite range*; it describes *local* correlations between the excited system particles. The presence of a term with an infinite range may be considered as characteristic of a *superfluid boson system.* In a normal system, there is no such term; the density matrix $\rho_o(\mathbf{r}, \mathbf{r}')$ vanishes as $(\mathbf{r} - \mathbf{r}') \to \infty$.

It is straightforward to extend these arguments to the case of a state $|\varphi\rangle$ other than the ground state. We assume that the state $|\varphi\rangle$ includes a *condensate*, containing a macroscopic number of particles. Among the intermediate states of (10.2), we again single out the state obtained by pulling one particle out of the condensate. Let us denote such a state as $|\varphi(N-1)\rangle$. The matrix element

$$\phi(r) = \langle \varphi(N-1) | \psi(r) | \varphi(N) \rangle \tag{10.10}$$

is again of order $\sqrt{N}$ (it involves the square root of the number of condensed particles). That single state will thus provide a contribution to the density matrix

$$\phi^*(\mathbf{r})\phi(\mathbf{r}') \tag{10.11}$$

which is of order N. In contrast, all the other intermediate states of (10.2) involve some sort of localized excitation, leading to a matrix element of order 1. Again, by summing over all "excited" states $|\varphi_n\rangle$, we obtain a contribution, $\bar{\rho}(\mathbf{r},\mathbf{r}')$, which is comparable to the first term (10.11), for small distances, but which vanishes when $(\mathbf{r}-\mathbf{r}') \to \infty$ because of destructive interference between the various terms.

In short, the density matrix of a superfluid Bose system in an arbitrary state $|\varphi\rangle$ may be written as

$$\rho(\mathbf{r},\mathbf{r}') = \phi^*(\mathbf{r})\phi(\mathbf{r}') + \bar{\rho}(\mathbf{r},\mathbf{r}') \tag{10.12}$$

where $\bar{\rho}(\mathbf{r},\mathbf{r}')$ vanishes when $(\mathbf{r}-\mathbf{r}') \to \infty$. The first term on the right-hand side of (10.12) reflects directly the existence of a condensate; its presence in $\rho(\mathbf{r},\mathbf{r}')$ may be considered as *the* essential feature of superfluid Bose liquids, one which persists in the most complicated situations. This compact definition of superfluidity in the most general case was first given by Penrose and Onsager (1956).

The quantity $\phi(\mathbf{r})$ provides a measure of the condensate structure at point $\mathbf{r}$: it may, in fact, be regarded as the *condensate wave function*. Mathematically, it is defined by (10.10). Since, as we have seen earlier, it is not necessary to keep track of the exact number of condensed particles, we can replace (10.10) by the simpler relation

$$\phi(\mathbf{r}) = \langle\psi(\mathbf{r})\rangle \tag{10.13}$$

it being understood that the "average value" corresponds to an off-diagonal matrix element between two states with different condensate population. Let us emphasize that the definition (10.13) is completely general: it remains valid even if $\phi(\mathbf{r})$ varies appreciably over atomic distances.

To illustrate the character of $\phi(\mathbf{r})$ we consider three simple examples, beginning with the ground state $|\varphi_o\rangle$. In this case, the condensate wave function is

$$\phi_o(\mathbf{r}) = \sqrt{n_o} \tag{10.14}$$

corresponding to a uniform, fixed condensate. The *norm* of ϕ provides the number of condensed particles (in this respect, the definition of the "condensate wave function" differs from that introduced in Chapter 5).

Next we consider a moving condensate, with a uniform density and a slowly varying velocity. The system is then described by the London wave function (5.14), which for convenience we reproduce here:

$$\phi(\mathbf{r}\cdots\mathbf{r}_n) = \exp\left[i\left\{\sum_{i=1}^{N} S(\mathbf{r}_i)\right\}\right]\phi_o(\mathbf{r}\cdots\mathbf{r}_n) \tag{5.14}$$

(where S is real). In order to evaluate the off-diagonal matrix element (10.10), we notice that the operator $\psi(\mathbf{r})$ acts to destroy a particle at point $\mathbf{r}$, say particle number 1 (that choice is of no physical consequence, since the wave function is symmetrized). It follows that

$$\psi(\mathbf{r})\exp\left[i\left\{\sum_{i=1}^{N} S(\mathbf{r}_i)\right\}\right] = \exp\left[i\left\{\sum_{i=2}^{N} S(\mathbf{r}_i)\right\}\right]\psi(\mathbf{r})e^{iS(\mathbf{r})} \tag{10.15}$$

The matrix element (10.10) may therefore be reduced to the form

$$\begin{aligned}\phi(\mathbf{r}) &= e^{iS(\mathbf{r})}\,\langle\varphi_o(N-1)\,|\psi(\mathbf{r})|\,\varphi_o(N)\rangle \\ &= e^{iS(\mathbf{r})}\phi_o(\mathbf{r}) = n_o e^{iS(\mathbf{r})}\end{aligned} \tag{10.16}$$

In this London state, the modulus of ϕ is unchanged with respect to that in the ground state. On the other hand, $\phi(\mathbf{r})$ has a space-dependent phase. According to (5.26), the gradient of the phase of the superfluid wave function serves to define the superfluid velocity $\mathbf{v}_s$ at point $\mathbf{r}$ (at least on a macroscopic scale).

Finally we consider the state, described by the "generalized" London wave function (5.18), with $S(\mathbf{r}) = 0$. The system wave function is thus equal to

$$|\varphi\rangle = \left(1 + \sum_{\mathbf{q}} \alpha_{\mathbf{q}}\rho_{\mathbf{q}}^{+}\right)|\varphi_o\rangle \tag{10.17}$$

Such a state is characterized by small density fluctuations whose amplitude $\delta\rho(\mathbf{r})$ is given by (5.28). We may expect the density of condensed particles at point $\mathbf{r}$, to fluctuate as well. By analogy with the ground state, such a *condensate density* may be defined as

$$\rho_c(\mathbf{r}) = |\phi(\mathbf{r})|^2 \tag{10.18}$$

We shall see that in general, the fraction of condensed particles need not be constant for the density fluctuation here considered, i.e.,

$$\frac{\delta\rho_c}{\rho_c} = \frac{2\delta\phi}{\phi} \neq \frac{\delta\rho}{\rho}$$

In order to calculate the fluctuation

$$\delta\phi(\mathbf{r}) = \phi(\mathbf{r}) - \sqrt{n_o}$$

we substitute the wave function (10.17) into (10.10). Up to first order in $\alpha_{\mathbf{q}}$, we find

$$\phi(\mathbf{r}) = \sqrt{n_o} + \sum_{\mathbf{q}} \alpha_{\mathbf{q}} \left\langle \varphi_o \left| \rho_{\mathbf{q}}^+ \psi(\mathbf{r}) + \psi(\mathbf{r}) \rho_{\mathbf{q}}^+ \right| \varphi_o \right\rangle$$

On making use of the expansion (10.3) and of translational invariance, we may write $\delta\phi(r)$ as

$$\delta\phi(\mathbf{r}) = \sum_{\mathbf{q}} \alpha_{\mathbf{q}} e^{i\mathbf{q}\cdot\mathbf{r}} \left\langle \varphi_o \left| \rho_{-\mathbf{q}} a_{\mathbf{q}} + a_{\mathbf{q}} \rho_{-\mathbf{q}} \right| \varphi_o \right\rangle \tag{10.19}$$

The condensate density fluctuation is then given by

$$\begin{aligned} \delta\rho_c(\mathbf{r}) &= 2\sqrt{n_o}\,\delta\phi(\mathbf{r}) \\ &= 2\sqrt{n_o} \sum_{\mathbf{q}} \alpha_{\mathbf{q}} e^{i\mathbf{q}\cdot\mathbf{r}} \left\langle \varphi_o \left| \rho_{-\mathbf{q}} a_{\mathbf{q}} + a_{\mathbf{q}} \rho_{-\mathbf{q}} \right| \varphi_o \right\rangle \end{aligned} \tag{10.20}$$

This result should be compared with the total density fluctuation, obtained from (5.28):

$$\delta\rho(r) = 2 \sum_{\mathbf{q}} \alpha_{\mathbf{q}} e^{i\mathbf{q}\cdot\mathbf{r}} \left\langle \varphi_o \left| \rho_{\mathbf{q}} \rho_{-\mathbf{q}} \right| \varphi_o \right\rangle \tag{10.21}$$

It follows that $\delta\rho$ and $\delta\rho_o$ are not generally related in any simple way.

In the case of a non-interacting Bose gas, $\delta\phi$ may be evaluated explicitly. The reader may easily verify that in such a case

$$\phi(\mathbf{r}) = \sqrt{n_o}\,[1 + \alpha(\mathbf{r})] \tag{10.22}$$

On comparing (10.20) with (5.29b), we see that

$$\rho_c(\mathbf{r}) = |\phi(\mathbf{r})|^2 = \rho(\mathbf{r}) \tag{10.23}$$

All the particles are thus condensed in the same state. By analogy with (10.16), we might expect the simple result (10.22) to remain valid for an interacting Bose liquid. Such an analogy is, in fact, wrong: (10.22), as well as (5.29b), is *false* in the case of an interacting system.

It is left as an exercise to the reader to calculate $\delta\rho$ and $\delta\rho_c$ in the case of a weakly interacting Bose gas, using the methods presented in Chapter 9.

It will be found that $\delta\rho_c = \delta\rho$, a result which is hardly surprising since to that degree of approximation the condensate depletion is negligible.

In conclusion, let us emphasize that the above examples of a condensate wave function have been chosen to illustrate the relationship between the present approach and that of Chapter 5. In practice, the concept of a condensate wave function is much more powerful; indeed, its major applications are to *microscopic* perturbations, for which the formulation of Chapter 5 is not usable. The preceding discussion thus should not hide the complete generality of the definition (10.13).

10.2 Dynamics of Condensate Motion

We consider now the time-dependence of the condensate wave function, $\phi(\mathbf{r},t)$. The simplest case corresponds to a stationary state in which the condensate structure does not vary with time.

For this case, one has

$$\phi(\mathbf{r},t) = \phi(\mathbf{r})e^{i\mu t} \tag{10.24}$$

$\phi(\mathbf{r})$ acts as a *parameter* of the system, in much the same way as did $\mathbf{q}$ in Chapter 5 or n_o in Chapter 9. For a given condensate structure, $\phi(\mathbf{r})$, one may construct a microscopic theory in which one calculates the elementary excitations, correlation functions, etc.

By using the Heisenberg representation, we may write the definition (10.10) of $\phi(\mathbf{r},t)$ in the form

$$\phi(\mathbf{r},t) = \langle \varphi(N-1)\left| e^{iHt}\psi(\mathbf{r})e^{-iHt}\right| \varphi(N)\rangle \tag{10.25}$$

A stationary solution, such as (10.24), corresponds to an *eigenstate* of H. For an N-particle system, that state has the energy $E(N)$. The exponent μ in (10.24) may thus be identified with the difference

$$E(N) - E(N-1)$$

μ is therefore the *chemical potential* in the state under consideration. The time dependence of the condensate wave function in a stationary state is thus very simple.

We next consider the problem of arbitrary condensate motion. One might hope that a description of such motion could be obtained via two distinct stages:

1. In the first stage, one would solve the dynamical equation for the condensate wave function $\phi(\mathbf{r},t)$.
2. Next, one would obtain the microscopic fluctuations appropriate to this condensate, (the elementary excitations, correlation functions, etc.).

In practice, the condensate motion is influenced by the degree of excitation of the system, so that the two stages of solution are not distinct. What is required is a *self-consistent* solution, which takes into account the coupling between the condensate and the elementary excitations. The problem thus becomes extremely complicated; it can only be formulated in the language of field theory in a way which lies well beyond the purview of this book. For that reason, we shall not attempt to set up a rigorous formulation of condensate motion. Instead, we shall write down an *approximate* equation of motion for the condensate. Such an equation, although it ignores many important features of the problem, displays the major physical aspects of condensate dynamics.

We first express the system Hamiltonian (9.1), in terms of the creation and destruction operators $\psi^*(\mathbf{r})$ and $\psi(\mathbf{r})$:

$$H = -\int d^3\mathbf{r}\psi^*(\mathbf{r})\frac{\nabla^2}{2m}\psi(\mathbf{r}) + \frac{1}{2}\int d^3\mathbf{r}\int d^3\mathbf{r}' V(\mathbf{r}-\mathbf{r}')\psi^*(\mathbf{r})\psi^*(\mathbf{r}')\psi(\mathbf{r}')\psi(\mathbf{r}) \tag{10.26}$$

In order to simplify the calculations which follow, we consider the special case in which the interaction $V(\mathbf{r}-\mathbf{r}')$ is a "contact interaction," of the form

$$V(\mathbf{r}-\mathbf{r}') = V_o\delta(\mathbf{r}-\mathbf{r}') \tag{10.27}$$

The Hamiltonian (10.26) thus becomes

$$H = \int d^3\mathbf{r}\left\{-\psi^*(\mathbf{r})\frac{\nabla^2}{2m}\psi(\mathbf{r}) + \frac{V_o}{2}\psi^*(\mathbf{r})\psi^*(\mathbf{r})\psi(\mathbf{r})\psi(\mathbf{r})\right\} \tag{10.28}$$

The equation of motion for the operator $\psi(\mathbf{r})$, expressed in the Heisenberg representation, is simply

$$i\frac{\partial\psi(\mathbf{r})}{\partial t} = [H, \psi(\mathbf{r})] \tag{10.29}$$

Let us replace H by its expression (10.28). We find

$$\frac{\partial\psi(\mathbf{r})}{\partial t} = -\frac{\nabla^2}{2m}\psi(\mathbf{r}) + V_o\psi^+(\mathbf{r})\psi(\mathbf{r})\psi(\mathbf{r}) \tag{10.30}$$

We now take the expectation value of the operator equation (10.30) in the state under consideration. On making use of the definition (10.13), we obtain

$$\frac{\partial\phi(\mathbf{r},t)}{\partial t} = -\frac{\nabla^2}{2m}\phi(\mathbf{r},t) + V_o\langle\psi^*(\mathbf{r})\psi(\mathbf{r})\psi(\mathbf{r})\rangle \tag{10.31}$$

(10.31) is the *dynamical equation* governing the time dependence of the condensate wave function $\phi(\mathbf{r},t)$.

According to (10.31), $\phi(\mathbf{r},t)$ is coupled to the expectation value of a product of *three* operators. We could write a similar equation of motion for this product, which would in turn be coupled to a product of five operators, etc. We would thus generate a full hierarchy of dynamical equations, and in so doing be led directly into the field theoretical formulation of the problem. Such mathematical complications may be avoided if we make the following approximation:

$$\langle\psi^*(\mathbf{r})\psi(\mathbf{r})\psi(\mathbf{r})\rangle \sim \langle\psi^*(\mathbf{r})\rangle\langle\psi(\mathbf{r})\rangle\langle\psi(\mathbf{r})\rangle \tag{10.32}$$

With the aid of such a *factorization approximation* , which was proposed by Gross (1961) and Pitaevskii (1961a), we find that (10.31) reduces to

$$i\frac{\partial\phi}{\partial t}(\mathbf{r},t) = -\frac{\nabla^2}{2m}\phi(\mathbf{r},t) + V_o|\phi|^2\phi(\mathbf{r},t) \tag{10.33}$$

The evolution in time of the condensate wave function, $\phi(\mathbf{r},t)$, is seen to be determined, in suitable approximation, by the non-linear differential equation, (10.33). Such an equation is far from simple; we shall devote the remainder of this chapter to considering the character of various possible solutions.

Before doing so, however, we comment briefly on the validity of the factorization approximation, (10.32). It is reminiscent of the "linearization" procedures employed to solve the equations of motion for the particle-hole pair excitations in the RPA; to that same extent it resembles the Bogoliubov approximation studied in Chapter 9. We may therefore conclude that it is likely valid as long as the particle-interactions are *weak* and the depletion of the condensate is small, that is, when

$$|\phi^2| = \rho_c \approx \rho \tag{10.34}$$

10.3 Simple Solutions of the Condensate Equation of Motion

Let us now inquire whether the wave functions we have proposed for various simple physical situations are, in fact, consistent with the non-linear condensate equation of motion, (10.33). We consider first the ground state, $|\varphi_o\rangle$. According to (10.14) and (10.24) the corresponding condensate wave function is:

$$\phi(\mathbf{r},t) = \sqrt{n_o}e^{-i\mu t} \tag{10.35}$$

If we replace ϕ by such an expression in the dynamical equation (10.33), we find

$$\mu = n_o V_o \tag{10.36}$$

which is identical to the result (9.25) obtained in Chapter 9 for a weakly-interacting Bose gas. Since (10.33) is valid only in that case, our different approaches are seen to be consistent. The result (10.36) may be considered as determining the *equilibrium* value of the condensate wave function ϕ as a function of the chemical potential μ.

The above argument is easily extended to the case of a fluid in uniform translation, moving at a constant velocity

$$\mathbf{v}_s = \frac{\hbar\mathbf{q}}{m}$$

The condensate wave function is then given by

$$\phi(\mathbf{r},t) = \sqrt{n_o}e^{i\mathbf{q}\cdot\mathbf{r}}e^{-i\mu t} \tag{10.37}$$

On making use of (10.33), we find that the chemical potential appropriate to such a state is

$$\mu = n_o V_o + \frac{1}{2}mv_s^2 \tag{10.38}$$

In fact, one may pass from (10.36) to (10.38) by simple Galilean invariance arguments.

Let us now turn to a slowly varying "generalized" London state. The corresponding condensate wave function may be written as

$$\phi(\mathbf{r},t) = e^{iS(\mathbf{r},t)}\left[1+\lambda(\mathbf{r},t)\right]e^{-i\mu_o t}\sqrt{n_o} \tag{10.39}$$

where μ_o is the chemical potential in the ground state, given by (10.36). We remind the reader that the phase factor e^{iS} describes superfluid motion, with a velocity $\mathbf{v}_s(\mathbf{r},t)$ given by (5.26)

$$\mathbf{v}_s(\mathbf{r},t) = \frac{1}{m}\text{grad } S(\mathbf{r},t) \tag{5.26}$$

while the bracket in (10.39) gives rise to small density fluctuations , with amplitude $\delta\rho(\mathbf{r},t)$. To the extent that (10.33) is applicable, the condition (10.34) must be satisfied: we have, therefore,

$$\delta\rho(\mathbf{r},t) \approx \delta\rho_c(\mathbf{r},t) \approx 2n_o\lambda(\mathbf{r},t) \tag{10.40}$$

Such a density fluctuation is associated with a shift $\delta\mu$ in the chemical potential, given by (5.73). On making use of the result (9.28), valid for a weakly-interacting system, we verify that

$$\delta\mu(\mathbf{r},t) = \frac{ms^2}{N}\delta\rho(\mathbf{r},t) \approx 2V_o n_o \lambda(\mathbf{r},t) \tag{10.41}$$

The physical meaning of the wave function (10.39) is thus quite clear.

We now substitute (10.39) into the dynamical equation (10.33). On making use of (10.36), and keeping only lowest-order terms in λ, we find

$$-\frac{\partial S}{\partial t} + \frac{i\partial\lambda}{\partial t} = \frac{(\mathbf{grad}\ S)^2}{2m} - i\frac{\Delta S}{2m} - i\frac{\mathbf{grad}\ \lambda \cdot \mathbf{grad}\ S}{m} - \frac{\Delta\lambda}{2m} + 2n_o V_o \lambda \tag{10.42}$$

(10.42) determines *completely* the motion of the condensate. To see this, we take the real and imaginary parts of (10.40): we obtain

$$\frac{\partial S}{\partial t} + \frac{(\mathbf{grad}\ S)^2}{2m} = -2n_o V_o \lambda + \frac{\Delta\lambda}{2m} \tag{10.43a}$$

$$\frac{\partial\lambda}{\partial t} + \frac{\Delta S}{2m} + \frac{\mathbf{grad}\ \lambda \cdot \mathbf{grad}\ S}{m} = 0 \tag{10.43b}$$

On making use of (5.26) and (10.40), we see that (10.43b) is simply the continuity equation for superfluid flow, (5.70),

$$\frac{\partial\rho}{\partial t} + \text{div}\ (\rho\mathbf{v}_s) = 0 \tag{10.44}$$

Next, let us suppose that λ varies sufficiently slowly that we may neglect $\Delta\lambda/2m$ in (10.43a). With the aid of (5.26) and (10.41), we may then cast (10.43a) in the form

$$\frac{\partial S}{\partial t} + \frac{1}{2}m\mathbf{v}_s^2 = -\delta\mu \tag{10.45}$$

On taking the gradient of (10.45), we recover the *dynamical equation* (5.85) for superfluid flow. The two fundamental laws of superfluid dynamics are thus contained in the single equation (10.42). We have verified that the equation of motion for the condensate, (10.33), provides a *complete* description of the problem.

10.4 Coherence Length

We now apply the condensate equation of motion, (10.33), to the problem of a spatially-varying condensate. This brings us naturally to the idea of a characteristic length over which such variations may take place. Such a *coherence length* has already appeared in the last example of the preceding section. There we saw that the last term of (10.43a) could be neglected as compared to $n_o V_o \lambda$ as long as the scale of variation of λ was larger than

$$\xi = (m n_o V_o)^{-1/2} \tag{10.46}$$

As soon as the *modulus* of ϕ (i.e., λ) varies appreciably over a distance ξ the macroscopic equation (5.85) is no longer applicable.

More generally, let us consider the original equation of motion (10.33). In a *stationary state*, for which $\phi(\mathbf{r}, t)$ has the form (10.24), (10.33) reduces to

$$\frac{-\nabla^2 \phi(\mathbf{r})}{2m} + V_o |\phi(\mathbf{r})|^2 \phi(\mathbf{r}) = \mu \phi(\mathbf{r}) \tag{10.47}$$

(10.47) is a non-linear eigenvalue equation, which is usually difficult to solve. Certain qualitative features of the solution are, however, readily apparent. Suppose μ is close to its equilibrium value, $n_o V_o$. In that case, the variation of ϕ in the vicinity of point $\mathbf{r}$ occurs on a scale

$$\left\{ m V_o \left(n_o - |\phi|^2 \right) \right\}^{-1/2} \tag{10.48}$$

Even if $|\phi|^2$ is appreciably smaller than n_o, the scale over which ϕ varies is at least of order ξ; ξ is the *minimum distance* over which the condensate structure may vary in a stationary state.

This conclusion is valid only if μ remains close to its equilibrium value $n_o V_o$. For instance, the solution (10.37), which describes a uniform translation, may display a variation of ϕ over distances much shorter than ξ: we simply have to consider q such that $q\xi \gg 1$. In such a case, the chemical potential (10.38) is much larger than $n_o V_o$: there is no contradiction with the above result. Put another way, the above discussion applies only to *fluctuations* in the condensate (which do not affect the average chemical potential). The length ξ is the *minimum scale of these fluctuations.* It may be considered as a *coherence length* of the interacting Bose liquid. We postpone for the moment the comparison of ξ with the characteristic lengths introduced earlier in Chapters 4, 5, and 9.

According to (10.46), $\xi \to \infty$ as the particle interaction vanishes. An ideal Bose gas thus has an infinite coherence length, in the sense that it cannot support stationary, localized fluctuations of the condensate. We have encountered such pathological behavior before, in Chapter 5; it is again directly traceable to the fact that the free Bose system has *no* compressibility. In contrast, an interacting Bose liquid can support stationary fluctuations, essentially because of the *stabilizing* action of particle interactions. The latter provide cohesion to the liquid, which acts to reduce the scale of any density fluctuation. In other words, the existence of a finite compressibility indicates the presence of a restoring force against any inhomogeneity. If the scale of the inhomogeneity is larger than ξ, one can achieve a "local equilibrium" corresponding to a stationary state.

On comparing (10.46) with (9.30), we see that ξ is likewise the wave length at which the density excitation spectrum passes from a phonon regime to a free particle regime. It therefore marks the *limit of a macroscopic description* [beyond which the dynamical equation (5.85) breaks down, the elementary excitations departing from a simple phonon law], as well as the *minimum scale over which the condensate structure may vary.* We note the importance of the finite compressibility in both cases.

We next inquire whether ξ, as presently defined, is related to the coherence length introduced in Chapters 4 and 5. In Chapter 4, ξ^{-1} was defined as the momentum at which multi-particle excitations begin to be appreciable; for $\mathbf{q} \sim \xi^{-1}$, the transverse current-current response function $\chi_\perp(\mathbf{q}, 0)$ departs from zero. In Chapter 5, ξ was defined as the minimum scale over which the London wave function is correct; below that scale the local correlations between particles are affected by the inhomogeneity of condensate motion. We have emphasized the relationship between these latter two definitions. In order to make it still more explicit we consider a small admixture of a phonon with wave vector $\mathbf{q}$ into the ground state, $|\varphi\rangle$. We may try to describe such a state by the London wave function

$$\left(1 + \alpha_{\mathbf{q}}\, \rho_{\mathbf{q}}^{+}\right) |\varphi_o\rangle \tag{10.49}$$

When $\mathbf{q}$ is too large, the London wave function breaks down because $\rho_{\mathbf{q}}^{+}|\varphi_o\rangle$ is no longer an eigenstate. But this transition marks precisely the onset of multi-particle matrix elements $\left(\rho_{\mathbf{q}}^{+}\right)_{no}$. The two coherence lengths introduced in Chapters 4 and 5 are thus intimately related (at least for a pure system).

Finally, we may identify the coherence length of Chapters 4 and 5 with that defined by (10.46). There are two general arguments for so doing. First, we note that multi-particle excitations cannot distinguish between a transverse and a longitudinal probe. Thus where they become important in determining the transverse response, they are likewise important in determining the longitudinal response; it follows that phonon-like behavior cannot be observed for values of q for which multi-particle excitations contribute to the transverse response functions. Second, we have seen that if we form a perturbation on a scale smaller than ξ, the condensate refuses to respond. One thereby changes the balance between the condensate and the virtually excited particles; the local correlations are modified and the London wave function is no longer valid.

The precise equivalence of the above definitions of ξ cannot be established in the framework of the present chapter, which is only concerned with the weakly interacting Bose liquid: in this case multi-particle excitations are altogether negligible. We shall thus limit ourselves to the above qualitative arguments. These arguments are in fact substantiated by the detailed macroscopic theory: it is found that for *a pure system*, all the above definitions of the coherence length are consistent; all yield a value close to (10.46). (For liquid ^{4}He, the coherence length is of the order of the roton wave length.)

10.5 Microscopic Structure of a Vortex Line

A natural application of the dynamic equation, (10.33), is to the study of the microscopic structure of a vortex line. A calculation of that structure has been carried out by Pitaevskii (1961a).[1] It shows clearly the way in which variation of ϕ over a coherence length eliminates the divergences which appear in the "semi-classical" treatment of vortex lines given in Chapter 5.

We consider a straight vortex line located on the z-axis, and use cylindrical co-ordinates (z, ρ, φ). For reasons of symmetry, the condensate wave function must take the form

$$\phi(\mathbf{r}) = f(\rho)e^{in\varphi} \tag{10.50}$$

In order for $\phi(\mathbf{r})$ to be single-valued, n must be an integer.

We may expect the solution to possess the following general features. Far from the vortex core (here the z-axis), $\phi(\mathbf{r})$ is slowly varying. The London wave function is then applicable; $f(\rho)$ will be a constant, equal

[1] We follow his treatment in this section.

to its equilibrium value $\sqrt{n_o}$. The velocity $\mathbf{v}_s$ at point $\mathbf{r}$ is then given by (5.26); it is equal to

$$\mathbf{v}_s(\mathbf{r}) = \frac{n\hbar}{m}\,\mathbf{grad}\,[\varphi] \tag{10.51}$$

and is thus tangential with a modulus $n\hbar/mr$. The integer n gives us the number of circulation quanta in the vortex. We shall consider only the case $n = 1$.

When $\rho \lesssim \xi$, the velocity $\mathbf{v}_s$ varies rapidly over a distance ξ. We then expect a change of $f(\rho)$, arising from the distortion of the condensate structure in the vicinity of the vortex core. In order to obtain the exact solution $f(\rho)$, we replace ϕ by its value (10.50) in the dynamical equation (10.47) appropriate to a stationary state. We thus find

$$\frac{1}{2m}\frac{d}{d\rho}\left[\rho\frac{df}{d\rho}\right] + \frac{1}{2m\rho^2}f + V_o f^3 = \mu f \tag{10.52}$$

The non-linear differential equation (10.52) has been solved numerically by Ginzburg and Pitaevskii. It is found that

$$f(r)\begin{cases} = n_o & \text{when } \rho \gg \xi \\ \cong n_o\dfrac{\rho}{\xi} & \text{when } \rho \ll \xi \end{cases} \tag{10.53}$$

where ξ is the coherence length (10.46). A plot of f is sketched in Fig. 10.1. As expected, the variation of f is spread over a distance ξ.

According to (10.53), the core of the vortex line has a finite radius ξ. Inside the core, $f(\rho)$ goes to zero, in such a way that $\phi(\mathbf{r})$ is *regular* on the

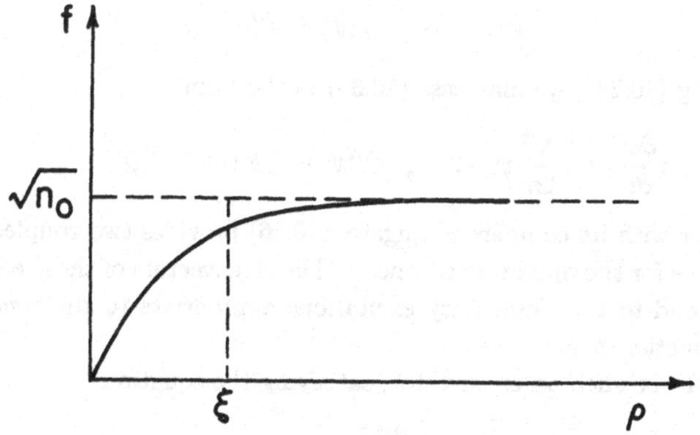

FIGURE 10.1. *Condensate behavior near core of vortex line.*

z-axis. The existence of a finite core serves to eliminate the divergences encountered in the macroscopic theory. For instance, the energy per unit length of a vortex line located along the axis of a cylindrical bucket (of radius R) is *exactly* given by

$$E + \frac{\pi N}{m} h^2 \log \frac{2.06R}{\xi} \tag{10.54}$$

The detailed solution of (10.52) eliminates the logarithmic uncertainty present in the former result (8.4).

Actually, the above treatment raises some difficulties, even for a weakly-interacting Bose gas. It is not clear to what extent the factorization (10.32) remains valid *in the vortex core*, since there ϕ may be much smaller than $\sqrt{n_o}$. However, although the precise solution $f(\rho)$ must be taken with some care, the approximate behavior (10.53) is certainly correct.

10.6 Elementary Excitations for a Non-Uniform Condensate

Once the condensate structure $\phi(\mathbf{r})$ has been found, we may in principle consider ϕ as a parameter, and search for the elementary excitations appropriate to the state with a vortex line. Such a calculation has been performed by Pitaevskii (1961a). Since it is mathematically somewhat complicated, we sketch only the method and the results.

Our starting point is the *operator* equation of motion (10.30). In the spirit of the factorization (10.32), we are led to replace any two factors in the product $\psi^*\psi\psi$ by their average values, the remaining factor being kept as an operator. We therefore write

$$\psi^*\psi\psi \to \langle\psi^*\rangle \langle\psi\rangle \psi + \langle\psi^2\rangle \psi^* \tag{10.55}$$

On using (10.24), we may cast (10.30) in the form

$$i\frac{\partial\psi}{\partial t} = -\frac{\nabla^2}{2m}\psi + 2V_o|\phi(\mathbf{r})|^2\psi + V_o\phi^2(\mathbf{r})e^{-2i\mu t}\psi^* \tag{10.56}$$

Together with its complex conjugate, (10.56) provides two coupled *linear* equations for the operators ψ^* and ψ. The eigenvectors of these equations correspond to the elementary excitations appropriate to the condensate wave function $\phi(\mathbf{r})$.

Let A^+ be such an eigenvector, satisfying the equation

$$i\frac{\partial A^+}{\partial t} = \varepsilon A^+ \tag{10.57}$$

By looking at (10.56), we see that A^+ must have the following form

$$A^+ = \int d^3\mathbf{r} \left\{ u(\mathbf{r})\psi^*(\mathbf{r})e^{i\mu t} + v(\mathbf{r})\psi^*(\mathbf{r})e^{+i\mu t} \right\} \tag{10.58}$$

where $u(\mathbf{r})$ and $v(\mathbf{r})$ are unknown functions. A^+ is the creation operator for the elementary excitation, ε its free energy. We note that (10.58) may be regarded as a generalization of the Bogoliubov transformation to the present non-uniform case. Let us substitute (10.58) into (10.57), and make use of (10.56). By collecting the coefficients of $\psi^*(\mathbf{r})$ and $\psi(\mathbf{r})$, we obtain two differential equations for u and v:

$$\begin{aligned} \varepsilon v &= -\nabla^2 v + \left\{ V_o|\phi(\mathbf{r})|^2 - \mu \right\} v + V_o\phi^2(\mathbf{r})u \\ \varepsilon u &= -\nabla^2 u + \left\{ V_o|\phi(\mathbf{r})|^2 - \mu \right\} u + V_o\phi^2(\mathbf{r})v \end{aligned} \tag{10.59}$$

The solution of (10.59) determines the elementary excitation spectrum.

In practice, one faces enormous mathematical difficulties, except in the case of the ground state, for which $\phi(\mathbf{r})$ is constant, equal to $\sqrt{n_o}$. In that case, the solution of (10.59) yields the excitation spectrum of the *weakly-interacting* Bose gas obtained in Chapter 9. The detailed calculation is left as a problem to the reader. For a liquid containing one vortex line, Pitaevskii has shown that there existed two types of excitations:

1. "Free" states corresponding to phonons or rotons *scattered* by the vortex line (in principle, one can use (10.59) to calculate the scattering cross-section, which determines *mutual friction*).
2. "Bound" states localized in the vicinity of the vortex line. Among these are found the *vortex oscillations* described in Chapter 8. Such oscillations thus appear as localized collective modes of the system.

The coupled equations, (10.59), though certainly complex in character, are rich in physical content. It may be expected that detailed study of these and similar equations will yield much new information concerning the non-uniform superfluid Bose liquid.

CHAPTER 11

CONCLUSION

We conclude our study of superfluidity in Bose liquids with a brief summary of the strengths, and of the inadequacies, of the theory. The cornerstone of our fundamental understanding of superfluidity has been the recognition of the central importance of the condensate, a single quantum state which is macroscopically occupied. For the Bose liquid one has a particle condensate, characterized by a non-vanishing expectation value of $\psi(\mathbf{r})$. The existence of the condensate explains all major aspects of superfluid behavior; in particular, we now possess:

1. A clear understanding of the static and the long wavelength dynamic properties of an equilibrium Bose liquid, including a microscopic basis for the two-fluid model.
2. A precise microscopic theory at $T = 0$ for a specific model, the dilute Bose gas.
3. Variational calculations of the excitation spectrum of liquid He II (and of the ground state properties as well), which are in good quantitative agreement with experiment.
4. A satisfactory phenomenological treatment of the interactions between elementary excitations which give rise to various irreversible phenomena.
5. A qualitative understanding of the role which is played by quantized vortices in both equilibrium and non-equilibrium phenomena.

Amongst the "lacunae" of the theory we may mention:

1. A microscopic theory of the ground state and elementary excitation spectrum of liquid He II; in fact, even the theory of the simple

dilute Bose gas has not yet been extended to treat the system at finite temperatures, and under non-equilibrium situations.

2. A clear understanding of a number of "vortex" properties, including the critical current for superfluid flow in thin capillaries.
3. A theory of the logarithmic singularity observed in the immediate vicinity of the λ-point.

Certainly successes of the theory outweigh its failures; equally clearly, a great deal of interesting work remains to be done.

REFERENCES

Allen, J. F. and Jones, H. (1938), *Nature* **141**, 243.
Allen, J. F. and Misener, A. D. (1938), *Nature* **141**, 75.
Allen, J. F., Peierls, R. F., and Udwin, M. Z. (1937) *Nature* **140**, 62.
Andreev, A. and Khalatnikov, I. M. (1963), *Sov. Phys. JETP* **17**, 1384.
Andronikashvili, E. L. (1946), *J. Phys. USSR* **10**, 201.
Atkins, K. R. (1959), "Liquid Helium," Cambridge Univ. Press, Cambridge, England.
Balian, R. and de Dominicis, C. T. (1971), *Annals of Physics* **62**, 229.
Beliaev, S. T. (1958), *Sov. Phys. JETP* **7**, 289, 299.
Bendt, P. J., Cowan, R. D., and Yarnell, J. L. (1959), *Phys. Rev.* **113**, 1386.
Bijl, A. (1940), *Physica* **8**, 655.
Blatt, J. M., Butler, S. T., and Schafroth, M. R. (1955), *Phys. Rev.* **100**, 476.
de Boer, J. (1963), in "Liquid Helium," ed. by G. Careri, Academic Press, N. Y., pp. 1–50.
Bogoliubov, N. N. (1947), *J. Phys. USSR* **11**, 23.
Brueckner, K. A. and Sawada, K. (1957), *Phys. Rev.* **106**, 1117.
Buckingham, M. J. and Fairbank, W. M. (1961), "Progress in Low Temperature Physics," Vol. 3, ed. by C. J. Gorter, North-Holland, Amsterdam, Chapter 3.
Chase, C. E. and Herlin, M. A. (1955), *Phys. Rev.* **97**, 1447.
Chester, G. V. (1963), in "Liquid Helium," ed by G. Careri, Academic Press, N. Y., pp. 51–93.
Cohen, M. and Feynman, R. P. (1957), *Phys. Rev.* **107**, 13.
Daunt, J. G. and Mendelssohn, K. (1939), *Nature* **143**, 719.

Feynman, R. P. (1955), in "Progress in Low Temperature Physics," Vol. 1, ed. by C. J. Gorter, North-Holland, Amsterdam, Chapter 2.
Feynman, R. P. (1964), *Phys. Rev.* **94**, 262.
Feynman, R. P. and Cohen, M. (1956), *Phys. Rev.* **102**, 1189.
Gavoret, J. and Nozières, P. (1963), *Annals of Physics* **28**, 349.
Ginzburg, V. L. and Pitaevskii, L. P. (1958), *Sov. Phys. JETP* **7**, 858.
Gross, E. P. (1961), *Nuovo Cimento* **20**, 454.
Hall, H. E. and Vinen, W. F. (1956), *Proc. Roy. Soc. A* **238**, 204, 215.
Henshaw, D. G. (1960), *Phys. Rev.* **119**, 9.
Henshaw, D. G. and Woods, A. D. B. (1961), *Phys. Rev.* **121**, 1266.
Hohenberg, P. C. and Martin, P. C. (1964), *Phys. Rev. Lett.* **12**, 69.
Hugenholtz, N. M. and Pines, D. (1959), *Phys. Rev.* **116**, 489.
Jackson, H. W. and Feenberg, E. (1962), *Revs. Mod. Phys.* **34**, 686.
Kamerlingh Onnes, H. (1911), *Proc. Roy. Acad. Amsterdam* **13**, 1903.
Kamerlingh Onnes, H. and Boks, J. D. A., *Leiden Commun.* **170a**.
Kapitza, P. L. (1938), *Nature* **141**, 74.
Kapitza, P. L. (1941), *J. Phys. USSR* **5**, 59.
Keesom, W. H. and Clausius, K. (1932), *Proc. Roy. Acad. Amsterdam* **35**, 307.
Keesom, W. H. and MacWood, J. E. (1938), *Physica* **5**, 737.
Khalatnikov, I. M. (1952), *JETP* **22**, 687; **23**, 8, 21, 169, 253, 265.
Khalatnikov, I. M. (1957), *Fortschr. d. Phys.* **5**, 211, 286.
Khalatnikov, I. M. (1965), "An Introduction to Superfluidity," W. A. Benjamin, N. Y.
de Klerk, D., Hudson, R. P., and Pellam, J. R. (1954), *Phys. Rev.* **93**, 28.
Kuper, C. G. (1955), *Proc. Roy. Roc.* **233**, 223.
Lamb, H. (1945), "Hydrodynamics," Dover, N. Y.
Landau, L. D. (1941), *J. Phys. USSR* **5**, 71.
Landau, L. D. (1947), *J. Phys. USSR* **11**, 91.
Landau, L. D. and Khalatnikov, I. M. (1949), *JETP* **19**, 637, 709.
Lane, C. T. (1962), "Superfluid Physics," McGraw-Hill, N. Y.
Lee, T. D., Huang, K., and Yang, C. N. (1957), *Phys.. Rev.* **106**, 1135.
Lee, T. D. and Yang, C. N. (1958), *Phys. Rev.* **112**, 1419.
Lifshitz, E. (1944), *J. Phys. USSR* **8**, 110.
London, F. (1938), *Nature* **141**, 643.
London, F. (1954), "Superfluids," Vol. 2, Wiley, N. Y.
Miller, A., Pines, D., and Nozières, P. (1962), *Phys. Rev.* **127**, 1452.
Onsager, L. (1949), *Nuovo Cimento* **6**, Suppl. 2, 249.
Palevsky, H., Otnes, K., and Larsson, K. E. (1958), *Phys. Rev.* **112**, 11.
Penrose, O. and Onsager, L. (1956), *Phys. Rev.* **104**, 576.

Peshkov, V. P. (1944), *J. Phys. USSR* **8**, 131.
Peshkov, V. P. (1946), *J. Phys. USSR* **10**, 389.
Pines, D. (1962), "The Many-Body Problem," W. A. Benjamin, N. Y.
Pines, D. (1963) in "Liquid Helium" ed. by G Careri, Academic Press, N. Y., pp. 147–187.
Pitaevskii, L. P. (1959), *Sov. Phys. JETP* **9**, 830.
Pitaevskii, L. P. (1961a), *Sov. Phys. JETP* **12**, 155.
Pitaevskii, L. P. (1961b), *Sov. Phys. JETP* **13**, 451.
Rayfield, G. W. and Reif, F. (1963), *Phys. Rev. Lett.* **11**, 35.
Sawada, K. (1959), *Phys. Rev.* **116**, 1344.
Tisza, L. (1938), *Nature* **141**, 913.
Tisza, L. (1940), *J. Phys. Radium* **1**, 165, 350.
Tisza, L. (1947), *Phys. Rev.* **72**, 838.
Vinen, W. F. (1958), *Nature* **181**, 1524.
Vinen, W. F. (1961a), in "Progress in Low Temperature Physics," Vol. 3, ed. by C. Gorter, North-Holland, Amsterdam, Chapter 1.
Vinen, W. F. (1961b), *Proc. Roy. Soc. A* **260**, 218.
Vinen, W. F. (1963), in "Liquid Helium," ed. by G. Careri, Academic Press, N. Y., pp. 336–355.
Ward, J. C. and Wilks, J. (1951), *Phil. Mag.* **42**, 314.
Ward, J. C. and Wilks, J. (1952), *Phil. Mag.* **43**, 48.
Whitney, W. V. and Chase, C. E. (1962), *Phys. Rev. Lett.* **9**, 243.
Wu, T. T. (1959), *Phys. Rev.* **115**, 491.
Yarnell, J. L., Arnold, G. P. Bendt, P. J., and Kerr, E. C. (1959), *Phys. Rev.* **113**, 1379.

INDEX

Printed in the United States
By Baker & Taylor Publisher Services

Printed in the United States
by Baker & Taylor Publisher Services